Multiple Regression
A Practical Introduction

Sara Miller McCune founded SAGE Publishing in 1965 to support the dissemination of usable knowledge and educate a global community. SAGE publishes more than 1000 journals and over 800 new books each year, spanning a wide range of subject areas. Our growing selection of library products includes archives, data, case studies and video. SAGE remains majority owned by our founder and after her lifetime will become owned by a charitable trust that secures the company's continued independence.

Los Angeles | London | New Delhi | Singapore | Washington DC | Melbourne

Multiple Regression
A Practical Introduction

Aki Roberts
University of Wisconsin-Milwaukee

John M. Roberts, Jr.
University of Wisconsin-Milwaukee

SAGE

Los Angeles | London | New Delhi
Singapore | Washington DC | Melbourne

SAGE

FOR INFORMATION:

SAGE Publications, Inc.
2455 Teller Road
Thousand Oaks, California 91320
E-mail: order@sagepub.com

SAGE Publications Ltd.
1 Oliver's Yard
55 City Road
London, EC1Y 1SP
United Kingdom

SAGE Publications India Pvt. Ltd.
B 1/I 1 Mohan Cooperative Industrial Area
Mathura Road, New Delhi 110 044
India

SAGE Publications Asia-Pacific Pte. Ltd.
18 Cross Street #10-10/11/12
China Square Central
Singapore 048423

Acquisitions Editor: Helen Salmon

Editorial Assistant: Kelsey Barkis

Production Editor: Gagan Mahindra

Copy Editor: diacriTech

Typesetter: diacriTech

Proofreader: Ellen Brink

Indexer diacriTech

Cover Designer: Lysa Becker

Marketing Manager: Shari Countryman

Printed in the United States of America

ISBN: 978-1-5443-3970-2

SUSTAINABLE FORESTRY INITIATIVE Certified Sourcing www.sfiprogram.org SFI-00756

This book is printed on acid-free paper.

20 21 22 23 24 10 9 8 7 6 5 4 3 2 1

• Brief Contents •

• Detailed Contents •

• Preface •

Introductory statistics courses are often required as part of the general education curriculum in colleges and universities, or as an element of major requirements in social science fields. After being exposed to introductory statistics, students may want to continue learning more advanced statistical methods, perhaps to enhance their marketability after graduation, as part of graduate study in their field, or out of sheer interest in the topic. But such students often find that the textbooks aimed at that next step are too technical in their presentation, frustrating those who are interested in learning more statistics but do not have an extensive background in mathematics. This is unfortunate, because with the great importance of quantitative methods in contemporary social science, and the role of multiple regression as a gateway to a variety of more advanced statistical techniques, the methods should be made available to students who might be stymied by a highly technical presentation.

Our goal in this book is to introduce multiple regression and related techniques in a way that is thorough yet still accessible to a wide range of readers. We aim to keep the presentation of this material as approachable and nontechnical as possible, with no expectation that readers have an extensive prior background in math or statistics. Although some use of equations and mathematical notation is unavoidable with this subject matter, we rely only on basic math that will be widely familiar to readers from all academic fields, and we give full discussions of any mathematical ideas that might not be so familiar. The statistical concepts presented here are all classic ideas that are discussed in other textbooks written at varying levels of mathematical sophistication, so the material is certainly not "new." But the specific emphases on different elements of the material here, the adaptations to a wide range of students' mathematical knowledge and confidence, and the particular ways in which the statistical ideas are conveyed have been carefully honed over our many years of teaching this content to undergraduate and graduate students in social science. The presentation will also work well for students in related fields that place a similar emphasis on quantitative data analysis.

Multiple regression asks how a *dependent* variable is related to, or predicted by, a set of *independent* variables. Such an analysis could, for example, involve data on a *sample* of American cities, in which the cities' crime rates (the dependent variable) are predicted by several of their economic and sociodemographic characteristics (the independent variables). A working knowledge of these tools is essential not only for those who aim to conduct such research themselves, but also for those who simply wish to read and appreciate the large body of existing research that utilizes these techniques and variations on them. We hope that studying the book and working through its examples and exercises will enhance readers' ability to carry out quantitative research of their own and also improve their understanding of research that others have done.

Our book consists of nine chapters, briefly described here. Chapter 1 contains background on quantitative research in the social sciences and provides a brief review of some core statistical ideas that are covered in standard introductory statistics courses and are needed before moving on to multiple regression. Chapter 2 focuses on the main elements of multiple regression and interpretation of results. The chapter also includes guidance on how to start a quantitative research project and how to organize a research paper that reports on multiple regression analysis. Chapter 3 extends the regression framework to include categorical independent variables such as race or religion, represented through dummy variables, and shows how results are interpreted in that case. Chapter 4 covers another widely used extension of multiple regression: interaction between independent variables. Interaction terms incorporate the idea that the effect of an independent variable on the dependent variable may vary depending on the value or level of another independent variable. For example, we may ask how the impact of education on income differs for women and men, or for people with different levels of work experience.

Chapter 5 discusses the use of logarithms and log-transformed variables in multiple regression. For a variable such as income, this will allow us to capture the intuitive idea that the same additive change, say a $5,000 increase, is more important when it occurs in the context of low rather than high incomes. The log transformation shifts our focus to multiplicative changes, such as a 10% increase. Chapter 6 frames the log transformation as an example of a nonlinear relationship between variables and describes how some other nonlinear relationships can be addressed with regression. We will be able to accommodate curvilinear relationships like that we would expect between age and physical strength, in which physical strength will typically rise with age up to a point, but then beyond that will usually decrease as age increases.

Chapters 1 through 6 are all concerned with the case of a numerical (continuous) dependent variable, but Chapters 7 and 8 introduce regression models for dependent variables that are categorical or counts. For categorical dependent variables such as employment status (employed or unemployed), we will focus on logistic regression and its extensions (Chapter 7). For count dependent variables, like the number of cars owned, we will use the framework of Poisson regression (Chapter 8). Finally, Chapter 9 gives brief overviews of several other regression-like methods that are popular in social science research; we hope that readers will take that chapter as an invitation to future study.

We anticipate that this book will be used in formal courses as well as in individual reading. Although the amount of material that can be comfortably covered in a course depends on many factors (whether the course is being given in a semester or quarter system, the number of classroom hours per week, whether there is a separate computer lab section, and so on), we expect that the first five chapters would be feasible to cover in a semester-long junior- or senior-level undergraduate course for social science students, especially if many of the Appendix sections are bypassed. An undergraduate course could also skip Chapter 4 (interaction) if needed to take a somewhat slower pace. A graduate course might move faster and thus be able to cover the entire book in a semester, or all except some sections omitted by the instructor. Students who are reading the book on their own as preparation for a more technically demanding multiple regression course can pick and choose topics as needed, perhaps in consultation with the instructor of their upcoming course.

The book includes many example analyses and interpretations, along with exercises for readers to work on as they proceed through the book. Each data set used for the examples and exercises is small enough for one to easily grasp the entire data set and its construction with respect to the specific statistical techniques covered in each chapter. Many of those data sets are "real" and come from various sources including the Census Bureau, the Uniform Crime Reports, the Centers for Disease Control, and the World Bank. The use of real data gives students some idea about where to find data if a term paper is required in the course. Other data sets are hypothetical, developed to help illustrate the statistical techniques used on them.

Statistical software of some sort will be required to do the exercises or reproduce the example analyses. Instructors using the book as a text can arrange their course to use whichever statistical software package they know best or is most accessible to students on their campus, and readers who are studying the book on their own can likewise use whichever software is most convenient. Fortunately, the methods covered here are all standard techniques that have been in use for many years and therefore will be available in any statistical software that is commonly used in the social sciences and related fields. For these standard methods, we expect that analysis of a given data set will yield essentially the same numerical results no matter which software package the analyst uses. Common software choices today include SPSS, Stata, SAS, and R, and a variety of older or more specialized statistical software packages would be fine too. Multiple regression can even be done in Excel, but we encourage readers to instead work with a statistical software package offering expanded capabilities that readers can access as their knowledge of quantitative methods grows. SPSS, Stata, SAS, and R guidance for analyses like those featured in the examples in the book is available on an accompanying website at **https://edge.sagepub.com/roberts1e**. so that readers can reproduce the examples' actual analyses with their chosen software. The website provides all example data in the various software formats. Instructors can also receive software guidance (for SPSS, Stata, SAS, and R), data, and solutions to all exercises.

In some places we mention that a particular idea is developed further in more advanced texts that are more rigorous mathematically, so that readers interested in more information can consult such texts. This is because some concepts in multiple regression and related methods really cannot be adequately conveyed without bringing in more technical material, and we would rather leave out some details and discussion of special cases or exceptions than risk overwhelming readers who are encountering this material for the first time. Readers who follow their study of this book by diving into those more advanced texts will encounter various intricacies that we have not examined in this book. But even then, it will be easier to tackle more technical material after gaining a practical understanding of regression here. Of course the presentation in the book is self-contained and does not require that any other texts be used alongside.

Students are sometimes surprised to learn that there can be heated controversies surrounding quantitative methods and their application in social science. Perhaps these students' intuition is that when a subject is rooted in math, the "right" and "wrong" approaches will be obviously clear to everyone. But actually there are substantial ongoing debates about many statistical and methodological issues, and collective understandings of the "correct" or "standard" way to carry out or interpret some analysis can change a good deal over time. It is common for even the most highly respected researchers to have their methodological decisions questioned, and commenters often challenge the methods by which a study addressed its research questions (see our discussion of openness to the scrutiny of others in Chapter 1).

We do not comment much on these controversies, as our goal is to provide readers with a solid understanding of multiple regression as usually applied by quantitative researchers in the social sciences, today and in recent decades. Of course we recognize that the "typical" quantitative analysis in social science could be much different in the future. For instance, the standard practice of most working researchers in the social sciences today relies heavily on the assessment of *statistical significance* of relationships between variables, and this idea plays an accordingly central role in our presentation of multiple regression. But this reliance on statistical significance has come under increasing criticism. Critics argue that statistical significance does not offer the best route to understanding the findings of a single study, and that, when used by researchers throughout a field, it also can distort the overall scientific record as documented in published research. It is possible that these criticisms could result in a marked shift away from reliance on statistical significance in coming years. So we do not claim that the picture of multiple regression painted in this book will forever reflect the typical practice of social science researchers, only that it does now and will probably continue to for some time. We hope that the foundational ideas presented here will leave readers ready to adapt to any future changes in the landscape of quantitative social science.

• Digital Resources •

Resources for both students and instructors are available to accompany this book at **https://edge.sagepub.com/roberts1e**.

For students: SPSS, Stata, SAS, and R guidance and data sets for the examples in the book, and solutions to two of the end-of-chapter exercise types: the interpretation of results, and questions about concepts. See the list of these exercises below.

Chapter 2: Exercises 2.6.2, 2.6.3
Chapter 3: Exercises 3.7.2, 3.7.3
Chapter 4: Exercises 4.6.2, 4.6.3, 4.6.4
Chapter 5: Exercise 5.6.1
Chapter 6: Exercise 6.4.1
Chapter 7: Exercises 7.5.2, 7.5.3
Chapter 8: Exercises 8.4.2, 8.4.3

For instructors: In addition to the resources outlined above, the instructor site will contain the data sets, software guidance for and solutions to a third exercise type: data analysis using statistical software and interpretation of the results. See the list of these exercises below. Also available on the instructor site are editable PowerPoints slides and a test bank.

Chapter 1: Exercise 1.4
Chapter 2: Exercise 2.6.1
Chapter 3: Exercise 3.7.1
Chapter 4: Exercise 4.6.1
Chapter 5: Exercise 5.6.2
Chapter 6: Exercise 6.4.2
Chapter 7: Exercise 7.5.1
Chapter 8: Exercise 8.4.1.

• About the Authors •

Aki Roberts is associate professor of sociology at the University of Wisconsin-Milwaukee. Her research interests include crime in the United States and Japan, crime clearance, and policing. She teaches undergraduate and graduate courses in quantitative methods and statistics.

John M. Roberts, Jr. is professor of sociology at the University of Wisconsin-Milwaukee. His research interests include social networks, quantitative methods, and crime. He teaches undergraduate and graduate courses on networks and other quantitative methods.

• Acknowledgments •

The beginnings of this book can be traced back to our earliest experiences with the course Sociology 481 at the University of New Mexico. John taught this course during his first year on the faculty there and decided to rely on class notes rather than assign any of the textbooks then available. In the years that followed, we each taught this course many times and steadily refined our presentation of multiple regression into what is now the material in the first part of the book. We thank our graduate student lab instructors and all of the students who have taken our Sociology 481 classes at the University of New Mexico and Sociology 461 classes at the University of Wisconsin-Milwaukee. Students' questions and feedback have greatly influenced our thinking about how to present this material. The second part of the book has also been informed by our teaching in Sociology 581 and 582 at UNM and Sociology 760 and 982 at UWM, and we thank those students too.

We are grateful to everyone at the Sage Publications, especially our editors Jeff Lasser and Helen Salmon. The book idea was developed with Jeff, and he worked with us through the first complete draft. Helen then guided us through subsequent revisions and publication. We also thank Kelly Huskey for taking the initiative to connect us with Jeff after Aki mentioned to her that we were thinking of writing a textbook. Michael Miner helped us too by providing Stata code for analyses that we originally carried out in SAS, SPSS, and R.

We also thank our families in Japan and the United States, especially Kazuko and Yoh Takeuchi, Rio Nishikawa Takeuchi, Tomoyuki Nishikawa, Tomohiro Nishikawa, Marilyn Roberts, James Roberts, Andrea Beauchamp, and Alexander Skutt for all their support.

Finally, we thank the late Raymond Liedka. Ray joined the faculty at the University of New Mexico shortly after John did, and had a keen interest in teaching quantitative methods to social science students at all levels. We enjoyed many discussions with Ray about the rewards and challenges of such courses, and we hope that this book would have earned his approval.

SAGE and the authors gratefully acknowledge feedback from the following reviewers during the development of this text:

- Joshua A. Goode, *University of Colorado Boulder*
- Jason Martin, *Eastern University*
- Margaret Ralston, *Mississippi State University*
- Catherine Harnois, *Wake Forest University*
- Juyeon Son, *University of Wisconsin Oshkosh*
- Jeannette Wade, *North Carolina A&T State University*

1 Introduction

This chapter provides background for the multiple regression methods that are presented in the remainder of the book. First, we discuss some of the potential advantages of quantitative research in the social sciences, along with some important cautions to heed. Second, we review some core statistical ideas, including descriptive statistics, hypothesis tests, correlation, and simple (bivariate) regression, which together form a foundation for understanding multiple regression. We illustrate those ideas with a full example and an exercise. Although we expect readers to have been exposed to these statistical concepts before, the material in this chapter will be a helpful refresher.

1.1 Quantitative Research

Social scientists engage in various kinds of research using different research methods, and the quantitative approach explored in this book does not characterize all of social science. However, this approach has proven helpful in investigating a wide variety of questions in social science and has been the main mode of inquiry in fields such as sociology, psychology, criminology, political science, and public health for many years. Before diving into the statistical material that makes up the remainder of this chapter, it is useful to discuss in brief some of the main attributes that make quantitative research approaches attractive as tools for social science research. Although this list is not exhaustive, we hope that it gives readers a sense of why the effort to learn these methods is worthwhile. We will also provide some cautions, as it is important to understand the limitations of these tools along with their strengths.

1.1.1 Benefits of Quantitative Research

One of the most appealing features of quantitative research is its openness to the scrutiny of others. When we disseminate a report or article describing a research study that we have conducted, critics can examine all parts of our work, including the decisions we made in preparing our data for analysis, our choices of analytic methods, and the interpretations we drew from our analytic results. They can attempt to verify our findings by repeating our analysis and confirming that we made no errors in applying the analytic methods that we used, or argue for alternative analyses and see if those lead to the same scientific conclusions about our research question. Critics can even look further back in the research process and challenge the methods that we, or others, used to collect data, or take issue with the overall design of the study. Even if this scrutiny does not suggest any problems with our study in isolation, other researchers can also check if our findings hold up when they analyze new data.

All of this can be uncomfortable for the researchers whose work is being examined; naturally no one enjoys being criticized, and it is hard not to take academic criticism personally. But this scrutiny is an essential part of an academic discipline's journey toward understanding, and a field's progress is held back if conclusions from flawed research are accepted as accurate. Note that this is not simply an issue for academic debates, because there can also be important consequences in the larger world when research is used to craft policy initiatives. Of course it is not always easy in practice for a field to carry out intensive scrutiny of the research that it produces. Researchers may resist requests to share their data with others, or research reports may not be clear enough for others to be sure as to precisely what methods were used to collect or analyze data. Still, even with these practical obstacles, quantitative research is more amenable to this scrutiny than are other types of research that do not rely on formalized procedures for collecting and analyzing data, or for which researchers do not have a well-defined data set that can be shared with others. Although there surely is also critical examination of academic work that relies on other kinds of research approaches, the nature of quantitative research makes this scrutiny especially possible and productive.

Along with facilitating critical examination of our research, quantitative methods can act as a check on ourselves when drawing conclusions from our research. Even if our intent is to be completely accurate and fair in our analysis and interpretations, human nature sometimes gets in the way. People are inclined to perceive patterns in data even when there really are not any, or put greater emphasis on observations that support what they already believe than on observations that contradict their beliefs. Although conscious misrepresentation of data or analytic results can occur, cases of that sort of intentional fraud or deception are likely rare. The bigger concern is that researchers who are honestly trying to be accurate and fair are still subject to these human flaws and biases.

Quantitative research, with its formal procedures for analyzing data and interpreting results, helps guard against these dangers. Patterns and relationships in data are deduced from formal analysis, rather than informal synthesis of observations, and researchers need to justify any decisions to exclude observations from their quantitative analysis. Of course researchers still need to make a wide variety of decisions when collecting data and choosing analytic methods, and some choices could potentially serve to push final results in a direction desired by the researchers. Likewise, interpretations of results are not always straightforward even when standard analytic procedures are used and therefore are subject to these dangers as well. So these threats cannot be completely eliminated even in carefully executed quantitative research. Nonetheless, quantitative research goes much further in addressing these concerns than research that essentially relies on the unaided human mind to carry out analysis and interpretation. We should be skeptical of such research no matter how intelligent the researchers are.

Another beneficial aspect of quantitative research is that quantitative measurement forces us to be as precise as possible about the social scientific concept that we intend to measure and how we can or cannot effectively measure it. Of course in any kind of empirical research we need to think carefully about the logic of research design, and all research likewise requires us to link broad concepts to the actual observations that are used in our study. Still, this linkage is especially salient in quantitative research, as we directly confront the issue of expressing theoretical concepts through concrete measurements. Some quantitative studies fall short in this regard, and the failure of a measure to adequately capture its intended concept is often a point of criticism when research projects are evaluated. But even if a measure is lacking in this way, the criticism will point to improvements that can be incorporated in further research. Formal measurement is therefore a key mechanism through which researchers can collectively work toward a better understanding of the world. When measurement is informal, it is much more difficult to even begin that conversation.

In practice, a good deal of social science research is done with at least some intent that the research will contribute to the development of effective public policy around some social problem or goal. The numerical findings that summarize the results of quantitative research can be especially helpful in this respect, as they can facilitate cost/benefit calculations that attempt to weigh the potential beneficial impact of a policy initiative against the expense required to implement it. For example, quantitative studies could help estimate the expected impact that a given increase in a school district's number of teachers may have on its high school graduation rate. Community members could then use this estimate as one element in deciding whether it will be worthwhile to direct more funds toward hiring teachers and, by implication, away from other programs supported by those funds. Even with quantitative estimates available, the ultimate decision will of course still require the school district to assess its priorities and make hard decisions about the services that it can realistically offer in light of its fiscal constraints, so the decision to add teachers surely involves much more than a single cost/benefit calculation. But relevant research results can be part of the debate in a much more helpful way than would be possible if the research findings could not be expressed numerically.

Another benefit of quantitative research is that it positions social scientists to take advantage of the ongoing development of new methods in the field of statistics. Even though there are important examples through the years of quantitative methods that were originally developed by social scientists, a substantial portion of the analytic methods that practicing social scientists use every day actually began in other academic fields, especially statistics. Statistical research focuses on developing and refining such methods and on understanding the conditions under which methods do or do not work as expected. By importing these methods and interpretations when required, social scientists can instead focus their intellectual energy on the formulation of social scientific research questions and how to add to our knowledge of the social world and improve our theories about it. Naturally a book like ours focuses on "workhorse" methods that have come to be the most widely used by social scientists over many years, rather than new methods that are at the cutting edge of statistical development. But the quantitative approach to research facilitates the eventual absorption of what are now cutting-edge methods into standard social scientific practice.

Finally, quantitative research is fun. Students often find that learning quantitative methods feels like coming upon an entirely new set of tools that can, if used appropriately, yield tremendous insight into many of the most important questions in social science. We hope that readers of this book will share this feeling of excitement.

1.1.2 Some Cautions

Quantitative methods offer tremendous promise, but researchers also need to be cautious in applying the methods and interpreting results. Some of these cautions will be mentioned in the context of specific methods introduced in later chapters, but it is helpful to highlight a few general issues here.

First, it is important to be clear that statistical analysis of quantitative data, even if highly sophisticated, technically advanced, and executed by knowledgeable researchers, typically cannot compensate for fundamental problems in the research design and data collection efforts underlying the data set being analyzed. In this book, we do not give much attention to all the parts of the research process that precede data analysis. But that is due only to limits of scope and space; careful attention to research design and data collection is absolutely crucial to carrying out successful and meaningful research. There are limited circumstances in which analytic methods have some ability to compensate for "bad" data after they have been collected, but such circumstances should be viewed as the exception, not the rule. The old adage "garbage in, garbage out" is usually pretty accurate. Although we focus here on quantitative analysis, in no way do we intend to minimize the vital importance of appropriate research design and data collection as discussed in social science research methods texts and courses.

A broad theme is that we need to guard against overinterpreting our research findings, or believing that they tell us more, and more definitively, than they really do. This general warning can take many specific forms. For instance, we must recognize that there will always be some uncertainty in the results that we report, at least if we intend to somehow generalize our findings beyond description of the particular sample used in our analysis. Some aspects of uncertainty will be addressed by statistics that we can report in our analysis, such as standard errors or hypothesis tests. But there is also an important sense in which a single study can rarely be truly conclusive, and a finding needs to be replicated in data from other samples or other contexts before we can begin to really trust it. Literal replication can be difficult for much research in the social sciences. Research may be situated in a historical setting that cannot be revisited, or in a *population* from which repeated sampling is unrealistic. But even if identical conditions cannot be repeated, and literal replication is impossible, we still will find research results more compelling if they seem to hold up in similar settings or populations. It is also crucial to acknowledge uncertainty in research findings when communicating with policymakers.

Another kind of overinterpretation is the *ecological fallacy*. As typically discussed in research methods texts and courses, the ecological fallacy refers to the distinction between the aggregate and individual *level of analysis*. The aggregate level refers to units like cities, states, or nations, representing collections of individuals and for which we can collect aggregated information like the percentage of the population that is unemployed or the average years of education for its adults. Individual-level units are, as the name suggests, the individuals themselves rather than collections of them. Then we could collect information about each individual, such as whether or not an individual is unemployed or the number of years of education that a person has completed. Distinctions between levels sometimes are not obvious. For instance, for some purposes a multiperson household may seem like an aggregate-level unit composed of a set of individuals, but for other purposes we envision a household making decisions and otherwise behaving as if it were an individual unit. For now, we will ignore this and suppose that there is a clear line between individual- and aggregate-level units.

On one hand, it is often easier for a researcher to obtain data on aggregate-level rather than individual-level units. Although many of the statistics that are reported for cities, counties, or states are ultimately based on surveys of individuals, or collected from official records that governments maintain on individuals, the aggregate-level information may be distributed in a way that makes it more accessible to researchers than the original individual-level survey data or records. And while data from many individual-level surveys are publicly available, a researcher may find that none of them include all of the desired information. (Because the same individuals generally do not appear in different surveys, it is usually not feasible to obtain data from more than one survey.) Sometimes individual-level measures can be obtained only by launching a new survey, but often that would be quite difficult and expensive. On the other hand, many of the theories and research questions that most interest social scientists actually lie at the individual level. For instance, the individual-level question of whether a person's employment status predicts their criminal behavior may seem more interesting than the aggregate-level analogue asking whether a city's unemployment rate predicts its crime rate.

The main objective is for the level of analysis implicit (or explicit) in the research question being studied to match the level of analysis at which the data used to address the research question were actually collected. That is, an individual-level research question should be considered in light of individual-level data. But noting the often greater accessibility of aggregate-level data alongside the attractiveness of individual-level research questions, it is not surprising that there is sometimes a mismatch. (The mismatch can be in the opposite direction too, with individual-level data being used to address an aggregate-level research question, but that is a less common problem.)

The ecological fallacy, then, refers to attempts to answer individual-level research questions via aggregate-level data. Using the example above, a finding of a relationship between city-level unemployment and crime rates cannot be taken as evidence that unemployed people are more likely to commit crime. It could be instead that within each city there is no pattern of unemployed people committing disproportionately many crimes, but rather that in places with higher unemployment, more crime is committed by *both* employed and unemployed people. Even if aggregate-level data should not be used to address individual-level questions or draw individual-level conclusions, researchers may find it very hard to resist doing so. The ecological fallacy is therefore a specific instance of the more general threat of overinterpretation of analytic results.

Naturally these few cautions do not exhaust the threats that one must be aware of when applying quantitative research methods. As we explore various specific techniques in the chapters that follow, we will sometimes warn of possible misinterpretations or common errors. These warnings should not dampen the excitement of working with quantitative research, but they do remind us that an analyst must always strive to exercise good judgment in using these tools.

1.2 Review of Basic Statistics

We conclude this chapter by reviewing some of the core ideas that are taught in typical beginning statistics courses for students of social science. We assume that readers have already been exposed to this material through a formal course or self-study. This brief review thus will be a refresher that leads into the new material covered in the rest of the book.

1.2.1 Descriptive Statistics

Descriptive statistics summarize the values of the variables for which information on the cases (the units of analysis) in a sample is available. The simplest description of a variable is the frequencies of the different values it takes in the sample. A *frequency distribution* or table typically reports the number and percentage (or relative frequency) of sample cases that take on each observed value of the variable.

Frequencies will be most appropriate for use with *categorical variables* (such as political affiliation) in which the values indicate category membership, or with discrete numerical variables for which there is a limited range of observed numerical values (for instance, the number of children that a person has). Frequencies can be impractical for use with continuous numerical variables (like height), which may take on very many different values among the sample's cases. Such a variable could even have a unique value for every case in the sample, especially if it is measured with great precision.

For a numerical variable that was measured on the sample cases, we can instead start by reporting the *mean* (usually denoted by a bar over the variable's label, such as $\overline{X}$) value of that variable. The mean is just a different name for the commonsense idea of an average: we simply add up the variable's values across all of the cases and divide by the number of cases N. The mean is interpreted as the typical value of X in the sample, so it is sometimes called a measure of central tendency. The *median* is an alternative measure of central tendency and is defined as the middle value after rewriting the sample values of X in order from lowest to highest. (If N is an odd number, it is clear which value is in the middle of the list; if N is even, one could take the average of the two middle values.) The median is especially appealing when the sample includes a small number of very high or low values that might distort the mean, but the mean is more important for the statistics that we discuss in this book.

Along with describing the typical value of a variable in a sample, it is important to also know how spread out the variable's values are. One way to measure this is the *range*, or simply the minimum and maximum values that the variable takes across the cases in the sample, along with the difference between those. The range is straightforward and informative, but one or two extreme values can of course have a big impact on it. The *variance* (usually denoted by s^2) is a more comprehensive measure of spread. It is based on the squared deviation from the mean of each case's value on the variable; these squared deviations are then summed over all cases and the sum is divided by N or (N – 1). For interpretation, the variance is usually transformed into the *standard deviation* (denoted by s) by taking its square root. The standard deviation can be interpreted as roughly the typical absolute (that is, ignoring positive and negative signs) deviation from the mean of the variable for cases in the sample.

1.2.2 Hypothesis Tests

In classical hypothesis testing, the analyst attempts to decide which of two distinct propositions, the *null hypothesis* (denoted H_0) and the *alternative hypothesis* (H_1), about some aspect of the population seems more plausible in light of the sample data. In this framework, the null hypothesis expresses current knowledge or belief about this population characteristic, while the alternative hypothesis is thought to hold if the null is not correct. (Note that in practice, the null may just be a convenient starting point for the research, rather than actually reflecting genuine knowledge or belief.) The analyst uses the sample data to decide whether or not to reject the null, where rejecting the null means that the alternative was found to be more plausible. One might see failure to reject the null as logically equivalent to "accepting" it, but researchers usually avoid that language; in principle, future research could lead to rejection of the null even if that did not occur in this study, so "accepting" seems too strong.

When the null indicates a particular value for some population characteristic, the *hypothesis test* is based on a comparison of the hypothesized value to its estimate in the sample. For example, if the null hypothesis is that the mean age in a population is 41 years, the mean age in the sample is compared to the hypothesized value 41. This comparison takes into account some assessment of sampling variability in the sample statistic; in the case of a single mean, the *standard error of the mean* represents sampling variability and is typically estimated by $s/\sqrt{N}$. A test statistic formalizes the comparison, using the form $\frac{\text{(Sample value} - \text{Hypothesized value)}}{\text{Standard error}}$. For a single mean, with the hypothesized population mean denoted by μ_0, this test statistic becomes $(\overline{X} - \mu_0)/(s/\sqrt{N})$.

If the null hypothesis is correct, statistical theory will indicate a particular probability distribution for the test statistic (in our example of a single mean, this will be a t-distribution). This allows calculation of a *p-value*, defined somewhat loosely as the probability, under the assumption that the null is correct, of obtaining a sample value that is as discrepant (or more) from the hypothesized value as was the statistic that we calculated in our sample. In the example of a single mean, suppose that $\mu_0 = 41$ and our sample yielded a mean $\overline{X} = 45$. Then the p-value, calculated by assuming that the population mean really is 41, reports the probability of obtaining a sample mean as far (or farther) from 41 as was our sample mean 45.

A small p-value pushes us toward rejecting H_0, because it suggests that either (a) the null really is true, but we happened by chance to get quite an unusual sample, or (b) our estimate in the sample just seems unusual because we calculated its probability under an incorrect premise in H_0. If the p-value is small enough, we will follow the logic of (b) and reject H_0, but we do that with a recognition that (a) could actually be what happened. The cutoff for deciding what is a "small" p-value is a statement of how concerned we are about incorrectly rejecting H_0 under scenario (a). The more worried we are about that possibility, the smaller the cutoff we would use, because a smaller cutoff leaves us less vulnerable to incorrectly rejecting H_0. In practice, there is a long tradition of using 0.05 as the cutoff for a "small" p-value, so that we reject H_0 if $p \leq 0.05$, and do not reject if $p > 0.05$. But it is important to recognize that the choice of 0.05 is essentially arbitrary, even though there is a long history of its use.

When H_0 indicates a particular value for the population (like $\mu_0 = 41$ in our example), we need to note whether the alternative is *one-sided* or *two-sided* (also known as one-tailed or two-tailed). This distinction refers to whether the alternative hypothesis proposes values that are only on one side of the hypothesized value, or on both sides. In our example of a single mean, a one-sided H_1 would indicate $\mu > 41$, or perhaps $\mu < 41$; either way, the alternative states that conceivable values of μ lie only above, or only below, the hypothesized value. A two-sided alternative accommodates both of these possibilities; in our example, this would be $\mu \neq 41$. The choice of a one-sided or two-sided H_1 does make a difference when calculating the p-value. The two-sided p-value is twice the one-sided p-value, because in the two-sided case we are asking about the probability that a sample statistic will be so discrepant from the hypothesized value in either direction. If we have enough understanding of the research topic to absolutely rule out the possibility of population values on one side of the hypothesized value, then it is good to use the one-sided alternative. But often we do not have such strong understanding that we can rule out these values, and the two-sided alternative would be appropriate.

Continuing with our example, suppose that we set up a two-sided alternative hypothesis, so that the null and alternative hypotheses were

- H_0: $\mu = 41$
- H_1: $\mu \neq 41$.

Suppose also that along with calculating $\overline{X} = 45$ from the data in a sample of 125 individuals, we calculated the standard error of $\overline{X}$ ($s/\sqrt{N}$ from above) to be 2.62. Then the test statistic is (45 – 41) / 2.62, or 1.53. We obtain the tail probability associated with the value 1.53 for a t-distribution with (125 – 1) = 124 *degrees of freedom* (df) from a table or online calculator. When using a table or online calculator, we need to note whether the alternative hypothesis is one-sided or two-sided. If we are using an online calculator that provides one-sided probabilities only, $t = 1.53$ corresponds to probability 0.064. Because H_1 was two-sided, we need to use both the upper and lower tails, which doubles the probability. Our p-value is therefore $0.064 \times 2 = 0.128$. However, we would obtain the value 0.128 directly if the online calculator provides two-sided probabilites. As this p-value is > 0.05, we do not reject H_0. Our conclusion is that the sample data are not so discrepant from the null hypothesis as to convince us that the null is implausible, and for now we believe that the hypothesis that the population's mean age is 41 remains reasonable.

When the null hypothesis is more complicated than simply indicating the value of a single population characteristic, the development of a test statistic and in turn the determination of a p-value will be more involved. But even if getting to the p-value is more complicated, the final step of comparing it to a cutoff (usually 0.05) to decide whether or not to reject H_0 will still apply. The chapters to follow will include various tests of hypotheses that are relevant in multiple regression analysis.

1.2.3 Correlation

Pearson's *correlation* or correlation coefficient (r_{xy}) allows us to assess the direction and strength (as defined below) of a relationship in our sample data between two numerical variables X and Y. (We do not explore any other types of correlation in this book, so for convenience we will drop the name "Pearson's" when discussing r_{xy}.) The direction of the relationship can be positive or negative, or there could be no relationship at all. In a positive relationship, the general tendency is for cases that have a high value of X to also have a high value of Y, and, conversely, for cases that have a low value of X to also have a low value of Y. High values of the two variables tend to be seen together, and likewise for low values. Figure 1.1 shows a scatterplot of values of X and Y for a hypothetical sample with 12 cases: a point is plotted at the (X, Y) values for the case it represents.

FIGURE 1.1 ● A Scatterplot of Values of X and Y (Positive Relationship)

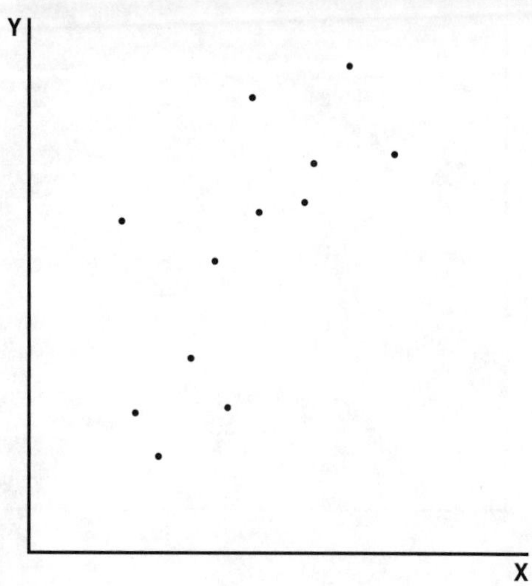

Figure 1.1 illustrates a positive relationship. Cases that have high values of X—toward the right side of the plot—also tend to have high values of Y, and cases with low values of X—toward the left of the plot—tend to have low values of Y. The cloud of points has a general shape that goes from lower left to upper right in the plot.

In a negative relationship, cases with low values of X tend to have high values of Y, and cases with high values of X tend to have low values of Y. That is, high and low values tend to be opposite for the two variables across cases. Figure 1.2 illustrates a negative relationship.

In Figure 1.2, cases on the left, with low values of X, tend to have high values of Y, and cases on the right, with high values of X, tend to have low values of Y. The cloud of points has a general shape of upper left to lower right. Compared with Figure 1.1, however, the pattern seems a bit less distinct here. For instance, the cases with the lowest values of Y in the figure are only average or slightly above average on X, rather than having especially high values of X. Still, the overall pattern is that of a negative relationship.

When there is no relationship between X and Y, the cloud of points does not exhibit either of the general shapes that indicates a positive or negative relationship. There is no tendency for cases with high (or low) values of X to have either high or low values of Y. Figure 1.3 shows a sample in which there is no apparent relationship between X and Y when we consider all points jointly.

As a result of the method by which the correlation coefficient r_{xy} is calculated, its value can range between –1 and 1. Its sign—positive or negative—indicates the direction of the relationship between X and Y. A positive value of r_{xy} corresponds to a positive relationship, and a negative value of r_{xy} corresponds to a negative relationship. If there is no relationship between X and Y, the value of r_{xy} will be zero.

FIGURE 1.2 ● A Scatterplot of Values of X and Y (Negative Relationship)

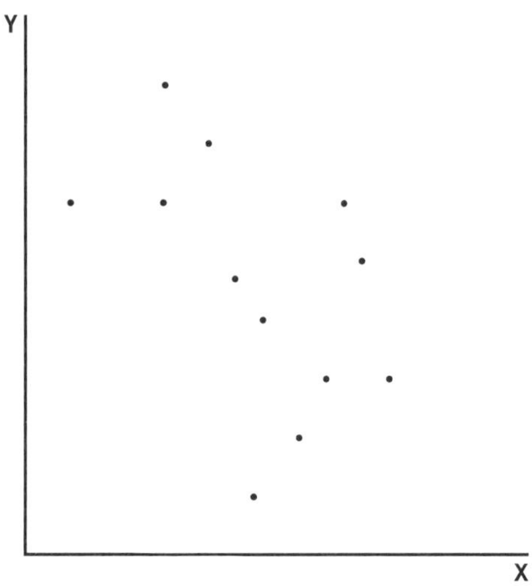

FIGURE 1.3 ● A Scatterplot of Values of X and Y (No Relationship)

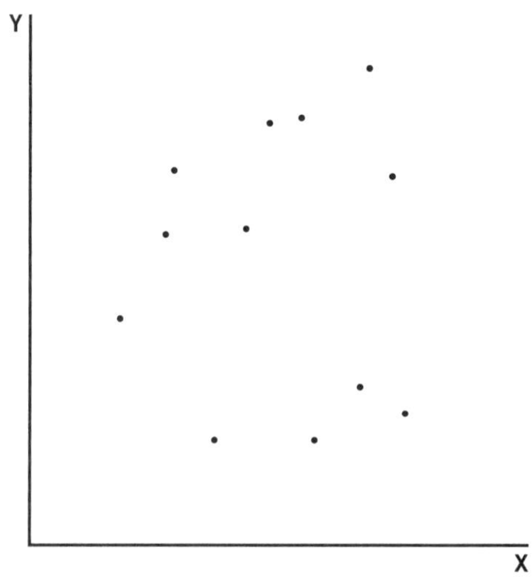

The magnitude of r_{xy} is also informative. It indicates the strength of the relationship between X and Y, in which a stronger relationship means that the points in the scatterplot are closer to lying on a single straight line. If the points are close to lying on a straight line, then for any particular value of X there is a very narrow range of Y values that appears along with it in the sample (and vice versa). In that situation, knowing what value a case had on one of the variables will give a very good idea of the likely values of the other variable. But if the points are far from lying on a straight line, then knowing the value of one of the variables is not so informative about the likely value of the other variable. In that situation the variables have a weaker relationship.

As mentioned above, r_{xy} can be no bigger than 1 and no smaller (more negative) than –1. Either of these extreme values indicates a perfect relationship between X and Y in the sample, with the points in the scatterplot exactly forming a straight line. (The perfect relationship is negative or positive depending on the sign of r_{xy}.) With a perfect correlation, one can precisely determine a case's value of one variable from its value on the other.

As r_{xy} gets closer to 0, indicating a weaker relationship between X and Y in the sample, the cloud of points looks less and less like a straight line. The sign still shows the direction of the relationship, but values close to 0 indicate a weak relationship, in which a case's value on one variable is not very informative about its value on the other. Figure 1.2 shows a somewhat less distinct relationship than Figure 1.1. That is borne out by the correlation coefficients for the data in the two figures. The data of Figure 1.1 result in $r_{xy} = 0.70$, suggesting a fairly strong positive relationship between X and Y. However, the data of Figure 1.2 result in $r_{xy} = -0.50$, a more moderate negative relationship. The points in Figure 1.1 are therefore closer to lying on a straight line than those in Figure 1.2. (The data of Figure 1.3 result in $r_{xy} = 0.00$, corresponding to no relationship.)

It is important to note that when we discuss the relationship between X and Y in terms of r_{xy}, we are actually referring only to a straight line, or *linear, relationship*. It is possible for a scatterplot to show a discernible pattern of how X and Y values go together in the sample but still give an r_{xy} close to 0 because that pattern is not consistent with a straight line. Usually a low r_{xy} genuinely does indicate no relationship between X and Y, but we should keep in mind that a more complicated *nonlinear relationship* could at least theoretically be present even when r_{xy} is close to 0.

1.2.4 Linear Regression

We can describe the relationship between X and Y further by performing a *simple* (or *bivariate*) *linear regression*. In the correlation, X and Y played equivalent roles, so we could haphazardly label one variable as X and the other as Y and it would not make a difference. But when we move to regression, we need to designate one variable as the dependent variable and one as the independent variable. The dependent variable is the outcome that we are attempting to explain or predict, while the independent variable is the factor that we believe predicts or, to be a bit bolder, determines the value of the dependent variable. The convention is that the dependent variable is labeled Y and the independent variable is labeled X. "Simple" linear regression refers to this situation of a single independent variable X influencing Y, but linear regression can be extended to include more than one independent variable, and that will be the focus of the rest of the book. For the moment, however, we will consider only analyses involving a single X. Our setup for the (simple) linear regression therefore can express a primitive theory in which X influences Y, even though we will often recognize that our research design does not allow us to convincingly say that the value of X truly "causes" the value of Y.

Linear regression fits a straight line (the *"regression line"*) to the cloud of points in the scatterplot. This line is the best (in a certain sense, to be discussed below) possible straight-line summary of the relationship between X and Y that is observed in the sample. In the discussion in Section 1.2.3, we saw that when r_{xy} is closer to 1 or –1, the summary provided by the line becomes a more accurate description of how X relates to Y in the sample data. That is because a high correlation indicates that the cloud of points is itself closer to being a straight line. As we mentioned with the correlation, the relationship between X and Y could be more complicated than can be represented by a straight line. Although linear regression inherently focuses on straight-line relationships, in later chapters we will see how the regression framework can actually accommodate various more complicated situations too.

Figures 1.4 and 1.5 show the same plots of 12 points displayed in Figures 1.1 and 1.2, but with the regression line added to each. In Figure 1.4, we see that the line has a positive *slope* (lower left to upper right), which agrees with our informal inspection of the cloud of points before, and is consistent with the sign of the correlation that we reported earlier. In Figure 1.5, the line has a negative slope (upper left to lower right).

FIGURE 1.4 ● Regression Line With a Positive Slope

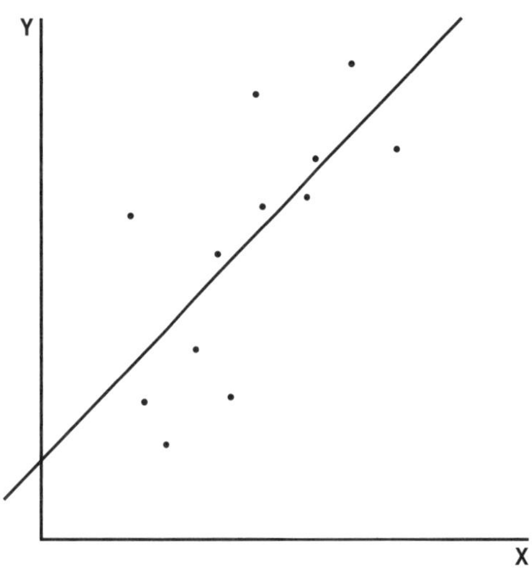

FIGURE 1.5 ● Regression Line With a Negative Slope

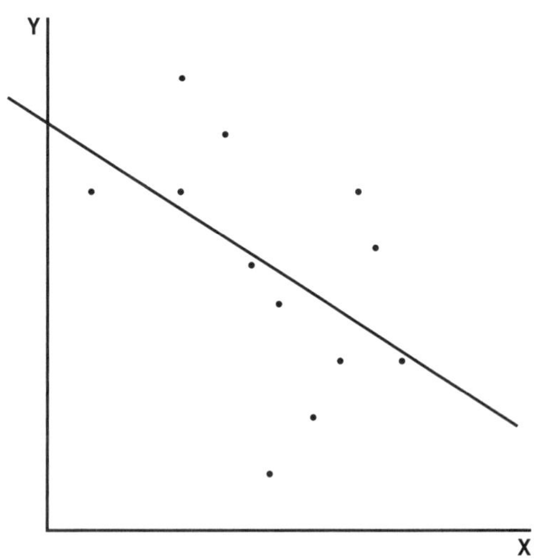

We can think of the line as representing a prediction of the value of Y for any value of X. That is, for any value of X that interests us, we can trace a vertical line from that place on the X-axis up or down to the regression line. The level of Y at which our vertical tracing meets the regression line is the predicted Y value for that X value. We can of course make such predictions for all of the cases in our sample, and see how the predicted Y for each case compares to its actual Y value. We call the differences between actual and predicted Y values the errors or *residuals*. In any sample there will be both positive and negative errors, unless r_{xy} is exactly 1 or –1 and all points are precisely on the line. The regression line is chosen to make these errors, as a group, as small as possible; that is the sense in which the line best represents the pattern of points. Because positive and negative errors would cancel each other out when summing the raw errors, the goal of making the errors as small as possible is interpreted in practice as making the *sum of squared errors* as small as possible. This is the *least squares* principle for choosing a regression line.

The line drawn on the scatterplot is informative, but for many purposes we need to instead work with the actual equation representing the line. The line is defined by its *slope* (the change in the predicted value of Y per unit change in X) and its *intercept*, the place at which it crosses the Y-axis, so those two numbers need to be indicated in the equation. We write the equation for the line in terms of predicted Y (or $\hat{Y}$, spoken as "Y hat," where the hat indicates the predicted value). The general form of the equation for the regression line with a single X is then $\hat{Y} = a + b X$, and the formulas that statistical software packages use to calculate numerical values of a and b from the sample data are determined by the least squares principle. Once we have numerical values for a and b, we can obtain the predicted value of Y for any X value by plugging that X value into the equation. For example, if our software calculated the regression equation for our sample data as $\hat{Y} = 5{,}320 + 300 X$, then the predicted Y when X = 10 is 5,320 + (300 × 10), or 5,320 + 3,000 = 8,320.

The slope b is the main object of interest in the regression equation, as it is directly providing information about the relationship between X and Y in our sample. That is, it tells how the predicted Y changes as we change the X value that we are plugging into the equation. Adding one unit to the plugged-in X value will change $\hat{Y}$ by b units: X is multiplied by b in the equation, and $b (X + 1) = b X + b$, so the calculated value of $\hat{Y}$ now differs by b units from what it was before adding 1 to the value of X. For example, if Y represents dollars of income and X represents years of age, b would tell us how the predicted dollars of income change when we plug in a value for age that is 1 year older than we plugged in before. For the regression equation above, in which b = 300, our summary of the information on X and Y in our sample data is that predicted income increases by $300 as the age we plug in increases by 1 year. If b were negative, then the change in $\hat{Y}$ would be a decrease. If b were –400, then the predicted income decreases by $400 as the age we plug in increases by 1 year. Because the regression equation refers to a straight line, this change in $\hat{Y}$ applies whether we are thinking about age 25 vs. age 26, or age 32 vs. 33, and so on. Note that when we refer to a one-unit increase in X, we mean this sort of interpretation of what happens when we plug an X value into the regression equation, not that we are altering our original sample data in some way.

In this way b gives us very specific information on how $\hat{Y}$ relates to X in the sample, and b is often described as the *effect* of X on $\hat{Y}$. The line is steeper when the absolute value of the slope b is larger, and flatter when the absolute value of b is smaller. If b = 0, then the line is perfectly flat, and our prediction of Y would be unaffected by the X value. In the equation, b = 0 would make the (b X) part of the equation zero no matter what X value we plugged in, leaving $\hat{Y} = a$ for any X value. That would occur when there is no relationship between X and Y in the sample. Although this language of "effect" is convenient, we should remember that it does not necessarily imply a true causal effect of X on Y.

In the social sciences, the intercept a in the regression equation is usually of much less interest to researchers than the slope b is. Of course the intercept must be included when calculating predicted values of Y, so it is certainly a necessary part of the equation. But it does not speak to the question of how Y relates to X like b does, and that relationship is usually the researchers' main concern. One specific interpretation of a is as the predicted Y when X = 0: if we plug in X = 0, then (b X) will be zero even if b is not, and we obtain $\hat{Y} = a$. In some cases it will be interesting to determine the predicted Y when X = 0, but in practice many variables used in social science do not have zero as a legitimate value. For instance, survey data on income will not include data on infants, and when X represents age, X = 0 will not occur in the sample data. Then even if the calculation is straightforward, there is not any meaningful reason to ask what the predicted income is for someone at X = 0. In any event, the direct information on the relationship between X and Y that is provided by b in the regression equation makes it our main focus for interpretation. Note too that in the regression output from most software, the column labeled "b" will include the value of a. a can be distinguished in the output by a term such as "intercept" or "constant" rather than a variable name.

We can calculate the regression equation no matter what the value of r_{xy} may be in our sample, but, as mentioned above, we regard the regression equation as a better representation of the sample data when r_{xy} is large (in absolute value). When r_{xy} is close to zero, the quality of the predictions for cases in the sample generally will be poor, and we will have many large errors or residuals relative to the overall variability in Y. In terms of the plot, there will be many points that have a large vertical distance from the regression line, so the regression line will not be a good fit to the cloud of points. Therefore, the value of r_{xy} is important context that we need to keep in mind when interpreting the specific information provided by the regression equation on the relationship between Y and X.

Note too that for most actual research questions, we will quickly recognize that surely more than just one independent variable would be helpful in predicting Y. In the example of age and income, we immediately think of education as another independent variable that would improve our ability to predict Y. Simple (bivariate) linear regression will therefore seem rather unrealistic in many research settings. Throughout the rest of the book we mainly focus on multiple regression, with more than one X included in the regression equation and hypothesized to have an effect on Y. In most social science applications, the richer multiple regression framework will be preferable to simple regression.

1.3 Example: State-Level Mental Stress in the United States

Throughout this book we have many fully worked-out and discussed examples that apply the methods to data, as well as many exercises meant to reinforce readers' understanding of the analytic techniques. We hope that readers will not only read the examples but also try to reproduce the examples' actual analyses as well as work through the exercises. Reproducing the examples or working on the exercises will require access to some statistical software. As noted in the Preface, any of the software packages that are commonly used in the social sciences R— for example, SPSS, Stata, SAS, or R— can be used for all of the analyses discussed in the book (and many more).

Our first example reviews descriptive statistics, correlation, and simple (or bivariate) regression. It involves a researcher who collected aggregate-level data on mental distress (the percentage of adults who reported that their mental health was not good for 14 or more of the past 30 days), median income (in thousands of dollars), and the percentage of single-parent households from the 50 U.S. states. She also categorized states into five geographical regions (MW, NE, SE, SW, and W). The state-level data collected from the Census Bureau and Centers for Disease Control are as follows:

Data:

State ID	State Name	Percentage Mental Distress	Median Income (in $1,000s)	Percentage Single-Parent Households	Region
1	Alabama	14.4	44.8	11.9	SE
2	Alaska	10.2	74.4	11.6	W
3	Arizona	11.7	51.3	11.6	SW
4	Arkansas	16.4	42.3	11.4	SE
5	California	10.6	63.8	11.5	W
6	Colorado	10.6	62.5	9.5	W
7	Connecticut	10.7	71.8	10.2	NE
8	Delaware	11.1	61.0	11.5	SE
9	Florida	11.4	48.9	10.7	SE
10	Georgia	12.6	51.0	13.0	SE
11	Hawaii	9.2	72.0	9.6	W
12	Idaho	10.7	49.2	9.4	W
13	Illinois	10.0	59.2	10.5	MW
14	Indiana	13.2	50.4	11.2	MW

(Continued)

State ID	State Name	Percentage Mental Distress	Median Income (in $1,000s)	Percentage Single-Parent Households	Region
15	Iowa	10.0	54.6	9.4	MW
16	Kansas	9.8	53.6	10.0	MW
17	Kentucky	14.7	44.8	11.3	SE
18	Louisiana	13.1	45.7	14.0	SE
19	Maine	12.6	50.8	9.2	NE
20	Maryland	10.1	76.1	11.6	SE
21	Massachusetts	11.9	71.0	9.7	NE
22	Michigan	13.4	50.8	10.6	MW
23	Minnesota	9.3	63.2	9.3	MW
24	Mississippi	14.1	40.5	15.2	SE
25	Missouri	13.2	49.6	10.6	MW
26	Montana	10.4	48.4	8.5	W
27	Nebraska	9.5	54.4	9.7	MW
28	Nevada	14.2	53.1	12.0	W
29	New Hampshire	12.7	68.5	8.7	NE
30	New Jersey	10.7	73.7	10.3	NE
31	New Mexico	12.5	45.7	12.9	SW
32	New York	10.6	60.7	11.1	NE
33	North Carolina	12.1	48.3	11.5	SE
34	North Dakota	9.0	59.1	8.4	MW
35	Ohio	12.9	50.7	11.2	MW
36	Oklahoma	14.3	48.0	11.6	SW
37	Oregon	13.0	53.3	9.6	W
38	Pennsylvania	12.6	54.9	9.8	NE
39	Rhode Island	13.5	58.4	11.3	NE
40	South Carolina	13.7	46.9	12.1	SE
41	South Dakota	8.3	52.1	10.1	MW
42	Tennessee	13.7	46.6	11.3	SE
43	Texas	10.6	54.7	12.8	SW
44	Utah	11.5	62.5	9.0	W

State ID	State Name	Percentage Mental Distress	Median Income (in $1,000s)	Percentage Single-Parent Households	Region
45	Vermont	11.9	56.1	8.8	NE
46	Virginia	10.9	66.1	10.2	SE
47	Washington	11.4	62.8	9.3	W
48	West Virginia	16.5	42.6	9.6	SE
49	Wisconsin	11.6	54.6	9.7	MW
50	Wyoming	12.1	59.1	9.3	W

After entering the above data in statistical software, the researcher ran the appropriate descriptive analyses for mental distress, median income (in thousands), percentage of single-parent households, and geographical region. Because mental distress, median income, and single-parent households are continuous numerical variables, the mean, median, standard deviation/variance, and range (minimum to maximum values) are all appropriate descriptive statistics. On the other hand, frequencies and percentages are appropriate descriptive statistics for region, because it is a categorical rather than numerical variable. Note that the squared standard deviations in the table do not quite match the corresponding variances, but that is simply due to rounding. Statistical software output will usually include more decimal places when giving these descriptive statistics.

	Mean	Median	Standard Deviation (Variance)	Range (Min-Max)
Percentage mental distress	11.9	11.8	1.8 (3.4)	8.2 (8.3–16.5)
Median income in $1,000s	55.7	54.0	9.2 (84.3)	35.6 (40.5–76.1)
Single-parent households	10.7	10.6	1.4 (2.1)	6.8 (8.4–15.2)

Region	Frequency (f)	Percentage (%)
MW	12	24.0
NE	9	18.0
SE	14	28.0
SW	4	8.0
W	11	22.0
Total	50	100.0

After running the descriptive statistics and becoming familiar with the data set, the researcher decided to explore the possibility that median income in a state influences, or at least predicts, the state's level of mental distress. In this case, median income is the independent variable (X) and mental distress is the dependent variable (Y). Correlation and (simple) linear regression are appropriate tools for examining the relationship between two numerical variables. The results of correlation and simple regression analysis are summarized below, in the output from statistical software:

Correlation Analysis

	Median Income in $1,000s (X)	Percentage Mental Distress (Y)
Median income in $1,000s (X)	1.000	−0.600
Percentage mental distress (Y)	−0.600	1.000

Simple linear Regression Analysis

	b	Standard Error of b	t	p-Value
Intercept (constant)	18.625	1.309	14.227	0.000 (<0.001)
Median income in $1,000s (X)	−0.121	0.023	−5.202	0.000 (<0.001)

The correlation between median income (X) and mental distress (Y) is −0.600, indicating a negative and moderately strong linear relationship between those two variables. The simple regression results show a value of 18.625 for the intercept a and −0.121 for the slope b. The regression equation for the relationship between median income (X) and mental distress (Y) in this sample is therefore $\hat{Y} = a + b\,X = 18.625 + (-0.121)\,X$. The b value shows that, from the data in this sample, predicted mental distress ($\hat{Y}$) will decrease by 0.121 units (0.121% of the state's adults) as the value of median income (in $1,000s) increases by one unit (or, said more naturally, the value of median income increases by $1,000). The negative value of b corresponds with the negative correlation, as both indicate a negative relationship between median income and states' levels of mental distress. Note that the software also produced several other statistics that for now we are not interpreting, including the *standard error of b*, t, and the p-value. We will explain those statistics in Chapter 2.

The researcher also believes that the percentage of single-parent households may predict the state's level of mental distress. To explore that, the percentage of single-parent households is set as the independent variable (X), and mental distress remains the dependent variable (Y).

Correlation Analysis

	Percentage Single-Parent Households (X)	Percentage Mental Distress (Y)
Percentage single-parent households (X)	1.000	0.379
Percentage mental distress (Y)	0.379	1.000

Simple Regression Analysis

	b	Standard Error of b	t	p-Value
Intercept (constant)	6.701	1.851	3.621	0.001
Percentage single-parent households (X)	0.488	0.172	2.836	0.007

The correlation between the percentage of single-parent households (X) and mental distress (Y) is 0.379, indicating that there is a positive but somewhat weak linear relationship between those two variables. According to the simple regression results, with intercept a = 6.701 and slope b = 0.488, the regression equation predicting mental distress is $\hat{Y} = a + b\,X = 6.701 + (0.488)\,X$. That is, the sample data can be summarized as indicating that a state's predicted level of mental distress ($\hat{Y}$) increases by 0.488 (0.488% of its adults) for each one unit (one percentage point) increase in the percentage of single-parent households.

1.4 Exercise: Suicide Rates in Japanese Prefectures

Japan is known for its high suicide rates compared with other industrialized countries. A government official wanted to investigate possible demographic and environmental factors related to suicide rates. He collected data on the suicide rate per 100,000 population, older adult (65 years and older) population percentage, and the average number of clear weather days per year for 47 Japanese prefectures (administrative units that are roughly like American states), which he also categorized by region. The data collected from the Japan Statistical Yearbook are as follows:

Prefecture ID	Prefecture Name	Region	Suicide	Older Adult Population	Clear Days per Year
1	Aichi	Chubu	19.35	20.13	23
2	Akita	Tohoku	32.97	29.47	9
3	Aomori	Tohoku	29.35	25.71	7
4	Chiba	Kanto	21.38	21.24	30

(*Continued*)

Prefecture ID	Prefecture Name	Region	Suicide	Older Adult Population	Clear Days per Year
5	Ehime	Shikoku	20.89	26.48	26
6	Fukui	Chubu	19.85	24.94	21
7	Fukuoka	Kyushu/ Okinawa	23.13	22.14	27
8	Fukushima	Tohoku	25.04	24.84	14
9	Gifu	Chubu	20.47	23.98	37
10	Gunma	Kanto	25.30	23.46	34
11	Hiroshima	Chugoku	21.22	23.66	25
12	Hokkaido	Hokkaido	25.30	24.66	8
13	Hyogo	Kansai	22.48	22.92	17
14	Ibaragi	Kanto	23.60	22.39	40
15	Ishikawa	Chubu	22.31	23.50	13
16	Iwate	Tohoku	32.03	27.07	8
17	Kagawa	Shikoku	21.59	25.40	25
18	Kagoshima	Kyushu/ Okinawa	24.27	26.38	32
19	Kanagawa	Kanto	20.89	20.11	26
20	Kochi	Shikoku	25.79	28.53	34
21	Kumamoto	Kyushu/ Okinawa	24.82	25.48	31
22	Kyoto	Kansai	22.15	22.99	18
23	Mie	Kansai	18.92	24.10	37
24	Miyagi	Tohoku	22.70	22.19	9
25	Miyazaki	Kyushu/ Okinawa	27.05	25.64	49
26	Nagano	Chubu	23.28	26.44	18
27	Nagasaki	Kyushu/ Okinawa	25.79	25.86	41
28	Nara	Kansai	19.06	23.84	25
29	Niigata	Chubu	28.43	26.16	9
30	Oita	Kyushu/ Okinawa	22.22	26.48	26
31	Okayama	Chugoku	20.67	24.94	29
32	Okinawa	Kyushu/ Okinawa	25.34	17.30	15

Prefecture ID	Prefecture Name	Region	Suicide	Older Adult Population	Clear Days per Year
33	Osaka	Kansai	23.64	22.14	18
34	Saga	Kyushu/ Okinawa	25.88	24.47	39
35	Saitama	Kanto	22.82	20.36	51
36	Shiga	Kansai	21.83	20.48	26
37	Shimane	Chugoku	25.66	28.87	17
38	Shizuoka	Chubu	22.68	23.69	42
39	Tochigi	Kanto	24.65	21.81	29
40	Tokushima	Shikoku	19.36	26.75	23
41	Tokyo	Kanto	21.48	20.08	31
42	Tottori	Chugoku	24.62	26.15	15
43	Toyama	Chubu	22.78	26.08	18
44	Wakayama	Kansai	24.85	27.05	25
45	Yamagata	Tohoku	26.26	27.54	8
46	Yamaguchi	Chugoku	24.05	27.91	28
47	Yamanashi	Chubu	27.00	24.57	33

1. After entering the above data in statistical software, use the software to calculate appropriate descriptive statistics for region, suicide rate, older adult population, and the number of clear days per year.
2. Use the software to perform correlation and simple linear regression analyses investigating the relationship between the older adult population and the suicide rate. Interpret the results as fully as possible.
3. Use the software to perform correlation and simple linear regression analyses investigating the relationship between the number of clear days per year and the suicide rate. Interpret the results as fully as possible.

2

Fundamentals of Multiple Regression

In this chapter, we present some basic ideas about *multiple*, or *multivariate*, regression analysis, including an introduction to multiple regression focusing on the difference between bivariate (simple) and multivariate regression, and interpretation of multiple regression results. We discuss predicting Y via a multiple regression equation and also the problem of *collinearity*. In addition, there are several examples of multiple regression analysis, as well as homework exercises. The chapter's Appendix A also provides guidance on how to start a research project involving multiple regression analysis, how to evaluate research hypotheses, and how to organize a quantitative research paper using multiple regression.

2.1 Introduction to Multiple Regression

Bivariate, or simple, regression examines the effect of an independent variable (X) on the dependent variable (Y). Multiple regression extends this idea by considering the effects of multiple independent variables (X's) on the dependent variable (Y). It is almost always more realistic for there to be multiple influences on a dependent variable than to suppose that truly only a single factor influences Y. For example, criminal behavior might be influenced by many factors such as economic hardship, lack of informal social control, and likelihood of getting caught and punished. Similarly, a person's income could be influenced by multiple factors such as age, gender, race/ethnicity, education, and work experience. In some analyses we cannot include all factors that might plausibly influence Y, as there are technical reasons for preferring that the number of X's be relatively small compared to the number of observations (N) in our data, or because we simply do not have all those measures available in our sample. Usually, though, this restriction is not very confining, and we are able to consider quite a few independent variables; we discuss the choice of variables later in this chapter.

Note that in both bivariate and multivariate regression, we decide whether an X influences Y, rather than the other way around. In making this decision we draw on theory, past research, or our common sense. We use our existing knowledge to decide whether to proceed as if "variable one" influences "variable two" or as if "variable two" influences "variable one." We label the variables X and Y based on this decision: in the first case, variable one is X and variable two is Y, while in the second case variable two is X and variable one is Y. Sometimes this can be quite difficult to decide, but in any case the direction of this influence will be assumed, not actually tested, in the methods we examine here.

Another point worth mentioning again is that we will often use the language of "effects" or "influence" of independent variables on the dependent variable even when, as noted in Chapter 1, the nature of the research design does not allow us to genuinely identify causal relations among the variables. Unless we are analyzing data from a true experiment, we are typically uncovering associations among variables rather than actual causal effects. Some formal methods attempt to draw *causal inferences* from nonexperimental data, but, except for a brief overview in Chapter 9 (Section 9.7, Causal Inference), those approaches are beyond the scope of this book. It is most convenient to simply speak of the effect of an X on Y, or X influencing Y, when discussing multiple regression results, but we should keep in mind that this is shorthand language, not necessarily an indication of a true causal relationship.

A single multivariate regression analysis includes multiple X's that might influence Y, and multiple regression aims to separate, or single out, the effect of each X on Y. Thus, the b (slope or *coefficient*) for a particular X in a multiple regression is interpreted as the effect of X on Y, expressed as how many units the prediction $\hat{Y}$ increases or decreases for each additional unit of X, while *holding other X's constant* or, in slightly different language, while *controlling for other X's*. The idea of "holding other X's constant" is the key conceptual element that distinguishes this interpretation of b from the interpretation made in bivariate regression. This is also why we cannot achieve the same results as a multiple regression by repeatedly applying bivariate regression (once for each independent variable). A series of bivariate regressions will not incorporate this idea of control/holding constant, so we need to consider all of the independent variables together in the same analysis.

In a bivariate regression, the apparent effect of X on Y may actually also incorporate effects of other X's that are related to it but are not included in the regression; although such a situation is also possible in multiple regression, it is especially likely in a bivariate regression that, by definition, includes only a single X. Suppose you are interested in finding the effect of age on income. It is easy to imagine that the variable "years in the labor force" also influences a person's income. Those two possible influences on income (age and years in labor force) are closely related conceptually, and will be correlated in any realistic data set: older people (those with higher ages) tend to also have more years in the labor force. Thus, the effect of age on income that is obtained in a bivariate regression probably also reflects, to some extent at least, the influence of years in the labor force.

A multiple regression can include both age (X_1) and years in the labor force (X_2) in one analysis, and attempt to statistically separate the effects of age (X_1) and years in the labor force (X_2) on income (Y). Thus, the b for age (X_1) from such a multiple regression analysis is interpreted as the effect of age (X_1) on predicted income ($\hat{Y}$), holding years in the labor force (X_2) constant (or, controlling for years in the labor force). In other words, we are trying to imagine what would happen if everyone in our data set had the same years in the labor force (X_2)—which of course will not be true in our actual sample—but varied in age. (If somehow everyone in the sample did have exactly the same years in the labor force, then we would not be able to include that as an X in our multiple regression. There must be at least some variation in an X for it to be useful in predicting Y.) In that imaginary scenario, how would differences in age (X_1) be reflected in differences in income (Y)? Similarly, the b for years in the labor force (X_2) from this multivariate regression analysis is interpreted as the effect of an additional year in the labor force (X_2) on predicted income ($\hat{Y}$), holding age (X_1) constant. (Note that we will use "effect of X on Y" and "effect of X on $\hat{Y}$" interchangeably; the first may sound more natural, while the second is arguably more precise.) That is, if we imagine that everyone in the sample had the same age (X_1), then b_2 indicates how differences in years in the labor force would be reflected in differences in predicted income (Y). This statistical separation of the independent variables becomes more difficult the more closely they are correlated; we will return to this concern later.

A multiple regression equation with three X's can be written in symbols as

$$\hat{Y} = a + b_1X_1 + b_2X_2 + b_3X_3, \text{ where}$$

- $\hat{Y}$ represents the predicted value of Y;
- a represents the Y-intercept;
- b_1 represents the effect (slope) of X_1 on Y, holding the other X's (X_2 and X_3) constant;
- b_2 represents the effect (slope) of X_2 on Y, holding the other X's (X_1 and X_3) constant; and
- b_3 represents the effect (slope) of X_3 on Y, holding the other X's (X_1 and X_2) constant.

We often refer to the regression equation as a regression *model*, because it embodies some social scientific hypotheses about which factors affect Y, and we recognize that as a model we do not necessarily expect it to capture all of the particular nuances of the data in our sample. In this book we will not focus on the technical details of actually calculating a and the b's from a particular data set; those details are covered in more advanced texts, and we will be relying on statistical software to do the calculations. It is enough to say that, as with simple regression in Chapter 1, these values are chosen under the least squares principle. That is, the values of a and the b's reported by our statistical software are chosen so as to make the predicted values $\hat{Y}$ in the sample as close as possible to the actual values Y, in the sense of minimizing the sum of squared errors (or residuals, meaning differences between the actual and predicted values of Y among the N observations in our sample). We will revisit the sum of squared errors in Chapter 3.

The above regression equation with three X's helps show why we interpret b_1 as the effect of X_1 on Y (or the change in $\hat{Y}$ as X_1 increases by one unit) while holding other X's constant. If we compare the predicted Y for a case calculated before and after increasing its value of X_1 by one unit, while holding its values of X_2 and X_3 constant, it is as if we changed the regression equation for that case from

$$\hat{Y} = a + b_1 X_1 + b_2 X_2 + b_3 X_3$$

to

$$\hat{Y} = a + b_1 (X_1 + 1) + b_2 X_2 + b_3 X_3, \text{ or } \hat{Y} = a + b_1 X_1 + b_1 + b_2 X_2 + b_3 X_3.$$

We can see, then, that the value of $\hat{Y}$ changed by b_1 as X_1 increased by one unit and other X's were held constant. This leads to the interpretation of b_1 as the effect of X_1 on Y while holding other X's constant. (Again, we will use the term "effect" for convenience even if our research design does not permit a genuinely causal interpretation.) If b_1 is a positive number, then $\hat{Y}$ increases when X_1 increases; if b_1 is negative, then $\hat{Y}$ decreases when X_1 increases. b_1, b_2, and b_3 are often called *regression coefficients*, because they multiply the values of the X's in the regression equation.

As in simple regression, we do not mean that we literally change our data by adding to the X values; our sample data do not change. Instead, we use this idea of a one-unit increase as a means of interpreting the results. Also as in simple regression, it may help to think of b_1 as the difference in $\hat{Y}$ between two cases that are equivalent except for a one-unit *difference* in their values of X_1. In many contexts this will seem more natural than thinking about a one-unit increase in X_1 for one case.

In a well-controlled lab experiment, the techniques of multiple regression would usually be less necessary. In that setting, it may be possible to focus on how changing a single factor affects Y, while controlling the research environment to such an extent that we are literally holding all the other influences on Y constant. For example, if we want to see how the amount of a specific chemical (X) in a solution influences the size of an explosion (Y), we can run the experiment with varying amounts of the chemical while, for instance, keeping the temperature, size of the container, and other important factors the same every time. Then, to the extent that we have successfully held these other factors constant, observed differences in the results (analogous to our Y) can logically be attributed to the changes that we made in the amount of the chemical.

However, in the social sciences there are a variety of practical and ethical reasons why it will be rare to have this degree of experimental control. Therefore multiple regression attempts to achieve statistically what we might be able to literally do in a science lab: distinguish the separate effects of different independent variables by seeing how Y changes when only a single X changes. The ability to do this when a lab experiment is impractical, or even logically impossible, and when the independent variables correlated with each other in real data, makes multiple regression extremely important in a wide range of social science fields. (We will also examine difficulties that can arise when X's are too highly correlated with each other; this is the problem of *collinearity*.)

2.2 Interpretation of Multiple Regression Results

We now know what multivariate regression is and how it differs from bivariate regression. In this section, we look at the core interpretations of multiple regression analysis results. We will need to pay attention to (a) *R-squared (R^2)*, (b) statistical significance (obtained from p-values), and (c) slopes, regression coefficients, or effects (b's). Note again that scientific conclusions drawn from the analysis of just one single data set are inherently somewhat tentative. We should keep that in mind as we focus on the technical interpretations of the results.

We should also be aware that researchers use a variety of terms to refer to the act of carrying out a multiple regression analysis. "Run a regression" is common but informal, while "estimate a regression model" highlights the fact that our analysis of the sample data is meant to estimate what we would find if we could actually analyze data on the entire population from which we drew the sample. (We discuss below how this sort of thinking can also apply to situations in which we have data on the whole population.) In any case, these different terms do not imply differences in the actual analysis being done.

2.2.1 R-squared—Overall Performance of Multiple Regression

One of the main purposes in carrying out a regression analysis is to predict values of Y. Therefore, it is important to know how well the regression is actually doing at predicting Y in our sample. How close are the predicted values of Y to the actual values for the sample cases? If the predicted values of Y closely match the actual values, the regression is performing well in one important respect, and the regression model has "good fit" to our data. If the predicted values of Y do not closely match the actual values, the regression is not performing well in this respect, and the regression model has "poor fit" to our data.

In bivariate regression, we can look at the scatterplot of X against Y and, more formally, the correlation r_{xy} between X and Y to indicate the fit of the regression. If points in the scatterplot are generally close to the bivariate regression line, then the bivariate regression model has a good fit to the sample data. Remember that the correlation between X and Y will be high (in absolute value) in this situation, so r_{xy} indicates the bivariate regression's fit. As r_{xy} gets closer to the extremes of 1 or –1, the fit of the regression gets better; as r_{xy} gets closer to 0, the fit of the regression gets worse. It also can be shown that the absolute value of r_{xy} is exactly equal to the correlation between Y and $\hat{Y}$ (from the bivariate regression), which is another justification for interpreting r_{xy} as a measure of how closely predicted and actual values of Y match in the sample. For example, if $r_{xy} = 0.90$, the bivariate regression has a very good fit: in that sample, predicted values of Y from the regression will be in general quite close to actual values of Y.

For multivariate regression, the situation is a little more complicated. Because we have to consider multiple X's together when determining the fit of the multiple regression model, the correlation between any single X and Y is not by itself adequate for assessing the fit. But we can still use the idea of the correlation between actual and predicted values of Y in the multiple regression context, with the only difference from the bivariate case being that the predicted values are based on the multiple regression. The square of this correlation between Y and $\hat{Y}$ is called "R-squared," written as R^2. It is the main indicator of the fit of a multiple regression model and is included in the regression output from any statistical software. R^2 will range between 0 and 1: it is the square of a correlation, so a squared value cannot be negative, and with a maximum possible correlation of 1 (in absolute value), the square cannot exceed 1 either. If R^2 is high (close to 1), then the multiple regression is predicting Y well, and the regression model has a good fit. If R^2 is low (close to 0), then the multiple regression is not predicting Y well and has a poor fit.

Poor fit likely means that the multiple regression model is missing important X's that are also related to Y and would help with predicting it. Therefore, a low R^2 leads us to think about the possibility of including additional (or different) X's in the regression model. Poor fit could also indicate that a straight-line regression does not effectively capture the relationship between the X's and Y, but we will wait until later chapters to consider that sort of situation.

It is difficult to set hard cutoffs for what is a high or low value of R^2, because the value of R^2 is affected by many aspects of the data we are analyzing. For instance, in general it will be easier to achieve a high R^2 in a small data set (that is, with a small N) than in a large data set. The same R^2 value may therefore give a different impression depending on the sample size. Still, when R^2 is not fairly close to 1, even in a large data set, we want to think about what important X's may be left out. Note that R^2 will always go up, at least a little, when we add any X's to the regression. But if the added X's do not really help much in predicting Y, the improvement in R^2 will be small.

R^2 is sometimes described as the proportion of variance (or variation) in Y that is explained, or accounted for, by the X's in the multiple regression. If, for example, the value of R^2 is 0.60, then we can say that 60% of the variance in Y is explained or accounted for by the set of X's that we included in the multiple regression. We can think of the variance of Y as referring to the overall pattern of cases in our sample with high or low values of Y, and the regression model for Y estimates how the cases' X values lead to these varying Y values. That means that the predicted values of Y reflect the X values through the regression equation. If the predictions generally match the actual Y values poorly, leading to a low R^2, then there is a good deal of case-to-case variability in Y that is not represented in, or accounted for by, the regression equation. On the other hand, generally close matches between predicted and actual values of Y—giving a high R^2—indicate that the regression equation does seem to represent the key sources of case-to-case variability in Y.

2.2.2 p-Values: Statistical Significance of Each X's Effect on Y

R^2 indicates the multiple regression's overall fit or performance in predicting Y in the sample data. The assessment of *statistical significance,* from p-values, is the first step in investigating the importance of each X's effect on Y (again, controlling for the other X's that also appear in the regression model and not necessarily meaning a genuinely causal effect; we also could refer here to $\hat{Y}$ instead of Y). In this section, we discuss how the general idea of statistical hypothesis testing applies to statistical significance in regression. As mentioned in the Preface, some aspects of the role of statistical significance in the interpretation of regression results, and social science more broadly, are becoming more controversial. However, here we simply present it as typically used by social scientists now and in the published literature from recent decades.

In adapting statistical hypothesis testing to the situation of multiple regression, we are interested in the question of what the effect of an X on Y would be in the entire population, rather than what effect we find in the particular sample we have. Of course, sometimes we actually do have data on an entire population, for instance when we are doing an aggregate-level analysis using data from all 50 states. In such a situation, we can still view the data as being like a sample from a theoretical "superpopulation" that reflects the many ways that history could have played out differently for these 50 states. Although this image can sometimes seem a bit farfetched, the practical result is that we will treat such data as if it were a sample from a large population.

If we had the entire population or superpopulation, we could find the *"true" effect* of an X on Y, holding other X's constant, and write it as β to distinguish it from b, the effect of this X on Y in our sample. That is, b is the estimate we obtain from our sample, while β is the value we would obtain if we could analyze data from the entire population. The true regression model would therefore be expressed with β's, and the estimated model would be expressed with b's. As usual in statistical hypothesis testing, we can frame our thinking in terms of two competing hypotheses about the population: the null hypothesis H_0 and the alternative hypothesis H_1.

The most common null hypothesis in regression analysis, and the one that is assumed in the output from standard software packages, is "H_0: X has no effect on Y in the *true model,*" or, stated in symbols, "H_0: $\beta = 0$." It is possible to consider other null hypotheses that specify some other value of β, and those can be used in the general hypothesis test format that was discussed in Chapter 1, but other null hypotheses are not so common in social science research. In practice, most applications use a two-sided alternative hypothesis, and this will generally be the default for statistical software's calculation of p-values. The alternative "H_1: X has some effect on Y in the true model," or equivalently "H_1: $\beta \neq 0$," will then be paired with the usual null, so that the typical hypothesis test is set up as

H_0: (null hypothesis): X has no true effect on Y, or $\beta = 0$;

H_1: (alternative hypothesis): X has some true effect on Y, or $\beta \neq 0$.

After setting up H_0 and H_1 this way, we need to calculate a test statistic—in this situation, it will be a *t-statistic*—from the sample information and use it to determine a p-value and decide whether we should reject or not reject H_0. Rejecting H_0 would mean that H_1, the hypothesis that X does have some effect on Y, appears more reasonable from our sample data, so in that case the sample leads us to believe that there really is some effect of X on Y in the entire population. If we do not reject H_0, then H_0—the hypothesis that X has no effect on Y—still appears plausible in light of the sample data. That is, if we do not reject H_0 we are saying that the sample information does not allow us to confidently conclude that there actually is an effect of X on Y in the population. Of course we must always remember that, in any particular analysis, it is possible that we reached an incorrect conclusion of rejecting H_0 or not, which is one reason why we are reluctant to rely too heavily on a single study of some research question.

The t-statistic follows the general format shown in Chapter 1 and can be calculated as follows:

$$t = \frac{b - \beta_0}{\text{standard error of b}}$$

The standard error of b is an estimate of sampling variability in b; that is, it gives us a sense of how much b might vary across samples that we could have drawn from the population. Because we are most commonly testing the null hypothesis that $\beta = 0$, the above equation usually can be rewritten:

$$\begin{aligned} t &= \frac{b - 0}{\text{standard error of b}} \\ &= \frac{b}{\text{standard error of b}} \end{aligned}$$

We can calculate this t-statistic for each of the X's in our regression and then find p-values for the t-statistics from a t-table or an online t-calculator. Either will require us to know the proper degrees of freedom (df), and for this test df are calculated from the sample size as df = N – (the number of X's in the regression) – 1. With a p-value (here two-sided) in hand, we can decide whether or not to reject H_0, using the traditional (though again ultimately rather arbitrary) p-value cutoff of 0.05. With this cutoff, we do not reject H_0 if the p-value is > 0.05. In that case we still believe that it is plausible that, in the entire population, X has no effect on Y, and we will say that "X does not have a statistically significant effect on Y." On the other hand, if the p-value is less than (or exactly equal to, which could occur due to rounding) 0.05, we reject H_0 in favor of H_1, believing that we would find an effect of X on Y if we were to analyze the whole population. In this case ($p \leq 0.05$), we say that "X has a statistically significant effect on Y." The language of statistical significance is therefore a summary of the results of this particular hypothesis test. When we are working with a multiple regression, it is good to also include "controlling for other X's" or "holding other X's constant" in our statement of the results.

Fortunately, much of the actual arithmetic here is done for us when we use statistical software, because the software will report each b's standard error and t-statistic, calculate the df, and find the p-value that corresponds to the value of the t-statistic. Our only work is to then use the p-values to decide whether each X has a statistically significant effect on Y. If the effect of an X on Y is statistically significant, then we will go on to interpret the slope b for that X. If the effect is not statistically significant, then we generally do not want to interpret b further, as we are not sure that there is any true effect of that X at all. Again, though, 0.05 is a rather arbitrary cutoff, so it is hard to justify why we would interpret b when the p-value is just under 0.05, but not when the p-value is just over 0.05. When the p-value is above but close to the cutoff, we may want to consider this an instance of "borderline," or "marginal," significance. We probably want to go ahead with interpreting b for an X that shows a borderline significant effect, but we need to make clear that in doing so we are not strictly applying the usual cutoff and that our conclusion is therefore even more tentative than usual.

2.2.3 Interpreting b in the Multiple Regression Context

When the p-value indicates that an X has a statistically significant effect on Y, we next want to interpret its slope b. b could be positive or negative, with a positive b indicating a positive relationship between that X and Y, controlling for other X's; as in the bivariate case, this means that X and Y tend to move in the same direction—as X increases, Y tends to also increase, and as X decreases, Y tends to also decrease—but now we also think of the other X's as being held constant when making this interpretation. Likewise, a negative b indicates a negative relationship between X and Y, controlling for other X's; as X increases (holding other X's constant), Y tends to decrease, and as X decreases, Y tends to increase. Also as in the bivariate case, this logic applies to comparisons of Y values for cases that differ in their X values, and this will often be a more helpful way to understand a positive or negative relationship between X and Y.

As discussed in Section 2.1, the specific numerical value of b indicates the effect of that X on $\hat{Y}$ (showing the increase in $\hat{Y}$ associated with an additional unit of X), controlling for, or holding constant, the other X's in the regression. Suppose that $b = 3$. Then we interpret the result as "$\hat{Y}$ increases by three units as X increases by one unit, controlling for other X's." If $b = -4$, we could try to use similar language: "$\hat{Y}$ increases by negative four units as X increases one unit, controlling for other X's." But this "negative increase" sounds very awkward. An increase of –4 is equivalent to a decrease of 4, so it is much more natural to say "$\hat{Y}$ decreases by four units as X increases by one unit, controlling for other X's." We can also think of b as reflecting the difference in $\hat{Y}$ between two cases that are identical in their values of other variables but differ by one unit on this X.

Let us consider a social science example. Suppose that we wanted to examine the influence of a person's education and criminal history on their income, with education measured by years of schooling (X_1), criminal history measured by lifetime number of arrests (X_2), and income measured as annual income in dollars (Y). We then collected information on these variables from a sample of adults and analyzed the resulting data.

Suppose that the multiple regression output from our statistical software indicated that the effects of both education (X_1) and arrests (X_2) on income (Y) are statistically significant ($p \leq 0.05$), with b_1 (the slope for X_1) = 2,580, and b_2 (the slope for X_2) = –5,890. Then our interpretation of b_1 is "predicted annual income ($\hat{Y}$) increases by \$2,580 as years of education (X_1) increase by 1 year, controlling for the number of arrests (X_2)." b_2 is negative, so we could say "For each additional arrest (X_2), predicted annual income ($\hat{Y}$) decreases by \$5,890, holding years of education (X_1) constant."

Although it is accurate to say "predicted" annual income as we have here, and helpful to focus attention on the fact that these figures are estimates from a model for the predicted Y ($\hat{Y}$), in practice we may assume that our audience understands that the b's refer to the predicted Y ($\hat{Y}$). If so, for convenience we could drop the "predicted" from this language, or use Y instead of $\hat{Y}$. Also, the data here came from a snapshot of the sample members at one point in time, rather than tracking individuals through time. Therefore this is an instance in which it may be more natural to view statements like "as education increases by 1 year, holding number of arrests constant" as referring to a comparison between people who differ by one year of education while having the same number of arrests.

2.2.4 Real-Life (Substantive) Significance

Real-life significance refers to an evaluation of whether the size of an X's effect on Y is large enough to be meaningful. That is, when considered in light of the definitions and observed values of the X and Y variables in our sample, does the effect seem important, or rather trivial? We always need to consider this question of real-life significance, because sometimes an X's effect is statistically significant (indicated by $p \leq 0.05$), but the size of the b is not large enough to suggest any meaningful impact on Y in real terms. Suppose that X_1 is years of education and Y is annual income, the p-value indicates that years of education has a statistically significant effect on Y, and b_1 is 25. This would mean that predicted annual income increases only \$25 for each additional year of education, holding other X's constant. Even 4 additional years of education would only increase predicted annual income by \$100: each additional year of education increases predicted income by \$25, so 4 additional years increase the predicted income by $4 \times 25 = \$100$. If the data are contemporary rather than historical, and so come from a context in which full-time workers have annual incomes in the tens of thousands of dollars, this result would suggest that, in practical terms, there is no meaningful impact of education on income. This tiny difference in predicted income between people differing by 4 years of education would mean that income is effectively unrelated to education. In that case, the effect of education on income would have statistical significance, but not real-life significance. Alternative terms for "*real-life*" here include *real-world,* practical, and *substantive.*

There is no automatic numerical cutoff for determining real-life significance. We might simply use our knowledge of social science, or our common sense, in deciding whether the effect of X on Y is large in real terms. However, it is better to examine descriptive statistics such as the mean, standard deviation, and range (minimum and maximum values) of the variables to determine whether the effect is large enough to have real-life significance. This requires examination of both X and Y. For X, it may be that the usual "one-unit change" is not the most suitable for this real-life assessment. For example, if X is public school spending per student, measured in dollars, a one-unit change in X refers to a single dollar. If the average public school spending per student in our sample were \$11,762, with a standard deviation of \$5,891, a minimum value of \$4,152, and a maximum value of \$32,366, a change of \$1 would appear extremely small. One additional dollar of spending per student will almost surely have virtually no effect on whatever Y we are studying, making it hard to assess real-life significance. In this case a larger change in X would be more interpretable, perhaps an increase of \$500 or even \$1,000 per student. Then the change in $\hat{Y}$ would be $500 \times b$, or $1{,}000 \times b$, and we will be in a better position to assess whether the resulting change in $\hat{Y}$ is large in real terms.

There could also be cases in which the usual one-unit change in X is too large. For example, the Gini index of income inequality in a society is usually presented as a number between 0 and 1. A "one-unit change" would be too large to sensibly consider for that variable, as it would be equivalent to a change from the theoretical minimum value to the theoretical maximum value of this index. Again the resulting change in $\hat{Y}$ would be calculated as b multiplied by the change in the Gini index, and we might consider something like a change of 0.10 instead of 1.

When we are convinced that we are examining an appropriately sized change in X, we next assess the resulting change in $\hat{Y}$, based on Y's descriptive statistics. To illustrate this process, suppose that, for aggregate-level data, X_1 is unemployment rate (percentage unemployed) and Y is the suicide rate per 100,000 population. Suppose too that, from our software's regression output, the p-value indicates that the unemployment rate has a statistically significant effect on suicide rate, and b_1 is 4. If our descriptive statistics for percentage unemployment (X_1) showed a mean of 5.5, a standard deviation of 0.9, and a minimum to maximum range of 3.8 to 6.9 in our sample, we would probably decide that a 1% change in unemployment is indeed appropriate for evaluating real-life significance. The value of b_1 means that the predicted suicide rate per 100,000 increases by 4 as the unemployment percentage increases by one unit (here, 1%). If suicide rates in the sample have mean 13.4 and range between 10 (minimum) and 80 (maximum), an increase of 4 in the suicide rate seems to be large enough to suggest real-life significance. On the other hand, the same estimated effect ($b_1 = 4$) may not seem to have real-life significance for a different dependent variable that had mean 1,200 and range 1,000 to 1,500.

We can revisit the earlier example of public school spending per student as X. Based on the descriptive statistics for public school spending per student (X) (mean \$11,762 with a standard deviation of \$5,891, minimum of \$4,152, and maximum of \$32,366), it seems that a \$1 increase in X is too small to be easily interpretable. We can instead consider the predicted change in Y when public school spending per student (X) increases by \$1,000. In this case, the change in $\hat{Y}$ is $b \times 1{,}000$. If b is 0.003 and Y is high school graduation rate (%), then we can say that the predicted high school graduation rate increases $0.003 \times 1{,}000 = 3\%$ as public school spending (X) increases by \$1,000, holding other independent variables constant. Then we can use the descriptive statistics for graduation rate (Y) to assess whether a 3% increase in graduation rate should be considered large enough to have real-life significance.

We always need to consider this question of real-life significance before deciding that a statistically significant X has a meaningful impact on Y, so we should not be too impressed by the words "statistically significant" alone. We still need to check real-life significance by looking at the size of b and the descriptive statistics for X and Y to reach a judgment as to whether the effect of this X is large enough to have some real impact. Of course different researchers may not always agree on whether the magnitude of a particular X's effect on Y is large enough to be called real-life significant, and there will be instances in which it is quite difficult to make this determination. Even so, attempts to assess the real size of any statistically significant effects are an important element in the full interpretation of regression results.

This is especially true when the sample size is very large. A large sample size (N) will tend to make the standard errors of b small, which will tend to make the t-statistics large and in turn tend to produce small p-values. It is, therefore, generally easier to find statistically significant effects of X's in large data sets, even if those X's actually have very small real-life impacts on Y. So it is especially important not to simply take statistical significance at face value when working with very large samples.

2.2.5 Other Notes on Interpretation

As in bivariate regression, we are usually not so interested in interpreting a, the Y-intercept. In the next section we look at calculation of the predicted value of Y, and a is certainly necessary for that. But in social science we rarely make a direct interpretation of a and usually do not discuss its value when interpreting results. We discuss this issue further in Section 2.3. Remember that, as we mentioned in Chapter 1, most software's regression output will provide the value of a in the "b" column, with a term like "intercept" or "constant" distinguishing a from the b's.

Note that along with "unstandardized coefficients," which are the b_1, b_2, and b_3 that we are discussing here, your statistical software may also present *"standardized coefficients."* Throughout this text, we focus on unstandardized coefficients, and b always refers to those. You may encounter standardized coefficients when reading journal articles or other research reports, however, and they can be thought of as giving another approach to the assessment of real-life significance. Therefore Appendix B to this chapter includes a brief discussion (see Section 2.8.1).

Research reports and articles sometimes present *confidence intervals* for regression coefficients. A confidence interval takes the general form of a sample estimate (of some population value) plus or minus a "margin of error" for that estimate. The confidence interval is reported with a numerical level of confidence, typically 95%. The interpretation of the 95% confidence interval is that the method for constructing the interval means that 95% of the time the interval will include the true population value. The analyst is therefore "95% confident" that the true value is somewhere in the interval.

A 95% confidence interval for a true regression coefficient like β_1 is found by $b_1 \pm$ margin of error. The margin of error reflects sampling variability in the estimate b_1, so it is based on the standard error of b_1. In particular, the margin of error is calculated by $t_{.025}$ (SE of b_1). The value of $t_{.025}$ is found in a t-table, using the usual df of t for assessing statistical significance of a regression coefficient (N – the number of X's – 1), and is the value that leaves a tail of the t-curve with 2.5% of the total probability. For instance, if df = 30, then $t_{.025} = 2.042$. Together, then, the 95% confidence interval for $\beta_1 = b_1 \pm t_{.025}$ (SE of b_1), or if written to highlight the two endpoints, as the interval ($b_1 - t_{.025}$ SE, $b_1 + t_{.025}$ SE). Note that $t_{.025}$ gets closer to 1.96 as df increase, so a quick approximation for typical sample sizes is that $t_{.025}$ is about 2, and the margin of error is roughly twice the standard error.

This confidence interval is closely related to the hypothesis test. If the β hypothesized by H_0 is within the 95% confidence interval, then the (two-sided) p-value for that H_0 is > 0.05. If that hypothesized value is outside the 95% confidence interval, then the p-value is < 0.05. (A value precisely on the boundary of the 95% confidence interval corresponds to p = 0.05, but that will be unlikely when results are given with several decimal places.) So for the usual regression situation in which H_0: β = 0, looking for 0 in the 95% confidence interval will give the same conclusion as assessing statistical significance in the usual way. But because the confidence interval does not require the notion of statistical significance or hypothesis testing to make sense, it may be appealing to researchers who are uncomfortable with the usual approach to statistical significance, even if it is closely related in a technical sense.

Confidence intervals are more popular in some fields than others, so we do not report them in the examples here. But the formula above is easy to execute, so confidence intervals can be constructed for any of the regression coefficients shown in the examples or exercises. The main difficulty with the confidence interval is that the commonsense interpretation that many people would make, that there is a "95% probability that β lies in the interval," is not really justified. In particular, the statistical framework that underlies the analyses here views β as fixed, even if it is unknown. That is, the framework does not include a notion of probability of different values of β; randomness enters via sampling variability in b_1, so we really want to say something like there is "a 95% probability that we obtain a value of b such that the resulting confidence interval includes β." This is certainly less intuitive, but the more natural statement will be sensible only if we have adopted a "Bayesian" framework in which there explicitly is a probability distribution for β. We do not explore that framework here.

2.3 Prediction in Multiple Regression

2.3.1 Calculating the Predicted Value of Y From the Values of X's

The process of obtaining predicted values of Y in multivariate regression is a straightforward extension of that in bivariate regression: to predict Y, we simply plug into the regression equation the values of the X's that we are interested in using. For example, suppose we wanted to investigate the effects of age (X_1), years of work experience (X_2), and years of education (X_3) on annual income (Y), and our analysis found a = –27,710, b_1 = –975.85, b_2 = 2,114.29, and b_3 = 5,580.44. Then the regression equation would be:

$$\hat{Y} = -27,710 - 975.85\,X_1 + 2,114.29\,X_2 + 5,580.44\,X_3$$

If we want to know the predicted annual income for a person who is 26 years old ($X_1 = 26$), has 4 years of work experience ($X_2 = 4$), and 16 years of education ($X_3 = 16$), we simply plug these values for X_1, X_2, and X_3 into the above regression and do the arithmetic:

$$\begin{aligned}\hat{Y} &= -27,710 - 975.85(26) + 2,114.29(4) + 5,580.44(16)\\ \hat{Y} &= -27,710 - 25,372.1 + 8,457.16 + 89,287.04\\ \hat{Y} &= 44,662.1\end{aligned}$$

Based on this regression equation, we can say that our analysis of the sample data indicates a predicted annual income of $44,662.10 for a person who is 26 years old with 4 years of work experience and 16 years of education.

Because this prediction uses estimates from the sample—the b's—rather than the true effects from the population—the β's—we could also think about sampling variability in the predicted annual incomes. That is, we could have a standard error for the prediction itself. Here we do not explore the calculations to estimate that standard error, but those can be found in more advanced texts.

2.3.2 Not Trusting the Results of Prediction

When we have a low R^2, we should not trust the prediction from the regression as much as we would when R^2 is high. Remember that a low R^2 indicates that the predicted values of Y are not in general very close to the actual values of Y for the cases in our sample, so a low R^2 will reduce our confidence in the quality of predictions in general. Also, we are uncomfortable with predicting Y based on a value of X that is well outside the range of X that was seen in the sample used to obtain the regression equation. Suppose that we are interested in the effect of age on some Y and that we obtained the regression equation from a sample in which all persons were 18 to 49 years old. Even if the resulting R^2 were high, we probably should not try to use the equation to predict Y for a 93-year-old person. As 93 is far outside the actual range of age in the data (18–49), we would not be too confident that the prediction for such a person is meaningful.

2.3.3 Why Not Interpret a?

We saw above that we use the value of a (the Y-intercept) in the arithmetic needed to predict Y. Earlier we said that we are rarely interested in directly interpreting a. We mentioned in Chapter 1 that in a bivariate regression, a can be interpreted as the predicted value of Y when X is 0; for multiple regression, a will be the predicted value of Y for the situation in which every X has value 0. If we plug 0 in for every X in the regression equation, the only non-zero part is a, so $\hat{Y} = a$. Social science applications seldom involve a situation in which it would be realistic for every numerical independent variable to equal 0. Then the scenario in which every X equals 0 is usually not very helpful to explore, and interpreting a as the predicted value of Y under this scenario usually will not lead to any scientific insight. (However, a situation like this can be interesting when all measures of independent variables are categorical; we discuss categorical independent variables in Chapter 3.)

2.4 Collinearity

With these fundamental interpretations of regression results in hand, we can begin to introduce some extensions of this core. *Collinearity* is, broadly speaking, a situation in which high negative or positive correlations among the independent variables inhibit our regression analysis. This does not require that all pairs of X's be highly correlated, so collinearity may be present if just some X or X's are highly correlated with one or more other X's. (There are also more complicated situations beyond high pairwise correlations in which collinearity can exist; advanced texts discuss this in detail.) Collinearity is potentially a concern because it causes increased uncertainty in the regression results.

In practice, one reason for highly correlated X's may be that a researcher is trying to measure one concept by multiple X's. For example, a survey respondent's social class background might be measured by father's and mother's education. In actual data, however, father's and mother's education are likely to be highly (and positively) correlated. We can imagine that people with a highly educated mother are likely to also have a highly educated father, as many couples meet in educational or occupational settings that tend to bring similarly educated people together. Or, at the aggregate level, consider the level of economic deprivation in a geographical unit like a city or census tract. Both the unit's poverty rate (perhaps measured as the percentage of households with incomes below an official poverty line) and its median income would be sensible measures of this concept. Again, though, these two measures are likely to be highly (and negatively) correlated: cities with a high poverty rate are likely to have low median income. Still, it is important to realize that there is not always an obvious theoretical reason for two independent variables to be highly correlated, so we will need to examine correlations among all the X's in our data, rather than just think about which pairs might be anticipated to be highly correlated.

When there is a strong correlation between two X's, it becomes more difficult to envision changing one while holding the other constant. More practically, it will be challenging for the regression analysis to determine the separate effects of two different X's if the cases in the data with a high value of one X almost always have a high (when there is a strong positive correlation) or low (strong negative correlation) value of the other. That is, if survey respondents with highly educated mothers tend to also have highly educated fathers, it will be hard to untangle whether mother's education, father's education, or both are actually influencing the dependent variable. In this situation, there will be more uncertainty in the estimated b's than there would be without such high correlations, and this uncertainty is reflected in larger standard errors. A tendency toward larger standard errors will mean a tendency toward smaller t-statistics (remember that the standard error is in the denominator of the formula for the t-statistics shown on the regression output: $t = \frac{b}{\text{standard error of b}}$), and in turn larger p-values.

Collinearity, then, will tend to make us less apt to find statistically significant effects of X's on Y. In the extreme, a classic symptom of collinearity is the seeming paradox of a high R^2, suggesting that the set of X's does quite well at predicting Y, yet with no or almost no statistically significant effects of the X's. This odd situation is possible under collinearity. The X's are together doing a good job of explaining the variance in Y, but high correlations among the X's mean that there is great uncertainty as to how, or which, particular independent variables are related to Y.

2.4.1 Diagnosing and Addressing Collinearity

The most basic approach to detecting collinearity begins by checking for any high positive or negative correlations among the X's. Although there are no strict cutoffs, correlations that are greater than about 0.70 in absolute value are especially apt to create problems in our regression. If two variables have an extremely high correlation (> 0.95, say) we surely will not want to include both in our regression. Otherwise, though, we can first run our analysis with all X's that we had originally identified as belonging in the regression, even those that are strongly correlated with each other. Perhaps the output will still show statistical significance for the effects of X's that are involved in the strong correlations, and we can conclude that these correlations are not really causing difficulties for our interpretation of results. But if some or all of the X's that are highly correlated do not show statistically significant effects, it may be that there is a collinearity problem.

It is not easy to "fix" collinearity. The simplest strategy is to identify pairs of highly correlated X's that did not exhibit statistical significance in the initial regression and then retain only one of each such pair when rerunning the regression. The choice of which of a pair to retain is somewhat arbitrary, as usually R^2 and the pattern of statistically significant effects will be quite similar for either choice. This is because two highly correlated X's are, in a sense, providing the same (or greatly overlapping) information in our analysis and will, therefore, lead to similar regression results. But if one of the pair seems most relevant to the theory and previous research that is guiding our analysis then that X can be chosen to remain in the regression.

There are other, more advanced methods for detecting and addressing collinearity; see Appendix B to this chapter for some discussion of these (Section 2.8.2). For beginning researchers, though, the simple strategy above is often quite effective. Of course even after removing X's from highly correlated pairs, some effects may not be statistically significant. If so, then the explanation for the nonsignificance is more likely the genuine absence of a relationship with Y, not collinearity. We clearly do not want to use collinearity as a catch-all "excuse" for nonsignificance. Usually nonsignificance really does mean no relationship between an X and Y, or at least not one that can be detected in the data we are analyzing. But in many instances, statistically significant effects will emerge once we address collinearity this way. When that happens, it is important to remember that our choice of which X to retain from a highly correlated pair was mostly arbitrary. Probably the other member of the pair would have given similar results had it been chosen instead, and we should keep that in mind when making interpretations. One or both of the pair are affecting Y, but we cannot really be more specific than that. In this situation, we should think of the X that we decided to include as a representative of this highly correlated pair, rather than thinking we have found that it is important for Y while the other X—the one we dropped due to the high correlation—is not.

2.5 Examples

The following examples illustrate the concepts we have discussed in this chapter.

2.5.1 Example 1: Crime in Colorado

A criminologist believed that economic deprivation, residential instability, racial inequality, and young male population have positive effects on the level of crime in society and wanted to test her research hypothesis with a multiple regression analysis. Economic deprivation, residential instability, racial inequality, and young male population were her independent variables, and the dependent variable was crime. Because her research question concerned aggregate-level relationships, she used aggregate-level data, in this case data from 24 cities in Colorado.

Economic deprivation was measured as the percentage of the labor force that was unemployed; a high percentage of unemployed persons in a city indicates a high level of economic deprivation. Residential instability was measured as the percentage of renter-occupied housing units; a high percentage of renter-occupied housing units indicates a high level of residential instability in the city. Racial inequality was measured by the index of white-Black residential segregation, with possible values from 0 to 100; a high value of the racial segregation index in a city indicates a high level of racial inequality. Young male population was measured as the percentage of the population that was male and aged 15 to 24 years. Crime was measured as the city's total violent crime (murder, robbery, rape, and assault) rate per 100,000 population (Y). She obtained these measures for each city from the U.S. Census Bureau and the FBI's *Uniform Crime Reports* (from the Bureau of Justice Statistics).

Her study can be summarized as follows:

Research hypothesis: Economic deprivation, residential instability, racial inequality, and young male population positively influence the level of crime in society.

Units of analysis: 24 cities in Colorado (aggregate-level data).

Measurements of independent and dependent variables:

X_1: Percentage of the labor force that was unemployed (economic deprivation).

X_2: Percentage of renter-occupied housing units (residential instability).

X_3: Racial segregation index (racial inequality).

X_4: Percentage of the population that was male, aged 15 to 24 years (young male population).

Y: Violent crime rate per 100,000 population (crime).

It is helpful to rewrite the research hypothesis using the actual measurements for the independent and dependent variables. We can also separate the different parts of the hypothesis.

Research hypothesis with measurements of independent and dependent variables:

1. The unemployment percentage (X_1) has a positive effect on the violent crime rate (Y).
2. The percentage of renter-occupied housing units (X_2) has a positive effect on the violent crime rate (Y).
3. The racial segregation index (X_3) has a positive effect on the violent crime rate (Y).
4. The percentage of young male population (X_4) has a positive effect on the violent crime rate (Y).

Her data are as follows:

City ID	City Name	Violent Crime (Y)	Unemployment (X_1)	Renter-occupied Housing (X_2)	Racial Segregation (X_3)	Young Male (X_4)
1	Arvada	125.9	7.3	26.7	24.3	6.1
2	Aurora	446.1	7.7	40.1	28.7	6.8
3	Boulder	211.6	7.0	52.3	20.3	16.7
4	Brighton	193.0	6.1	30.4	28.0	7.2
5	Colorado Springs	491.9	7.6	39.9	34.7	7.4
6	Commerce City	252.3	7.0	30.2	17.6	6.1
7	Denver	542.1	7.8	50.0	54.7	6.6
8	Durango	445.6	6.1	52.0	17.7	13.3

City ID	City Name	Violent Crime (Y)	Unemployment (X_1)	Renter-occupied Housing (X_2)	Racial Segregation (X_3)	Young Male (X_4)
9	Englewood	446.2	8.5	50.9	17.3	6.3
10	Evans	171.2	6.6	39.5	12.8	8.5
11	Federal Heights	478.8	9.4	48.0	20.2	7.5
12	Fort Collins	315.8	7.4	44.9	19.0	12.4
13	Fort Morgan	233.9	8.7	39.3	22.7	7.2
14	Golden	116.1	7.6	41.6	36.2	16.8
15	Grand Junction	349.0	5.4	37.6	22.4	7.9
16	Lafayette	153.5	7.2	27.2	13.4	5.7
17	Lakewood	442.7	7.8	41.1	24.4	6.7
18	Littleton	130.7	6.6	38.1	22.5	6.1
19	Longmont	311.5	7.0	36.5	10.2	6.3
20	Loveland	191.5	7.1	34.1	12.4	6.0
21	Northglenn	255.3	10.5	41.5	19.7	7.3
22	Pueblo	854.2	11.1	39.8	22.0	7.5
23	Sterling	487.3	7.7	41.1	63.8	10.7
24	Wheat Ridge	565.8	6.4	45.4	20.8	5.2

Year of data: 2010
Data sources: U.S. Census Bureau and Bureau of Justice Statistics, FBI's *Uniform Crime Reports*

After entering the above data in statistical software, the researcher checked correlations among the X's. The results of this correlation analysis are below:

	Unemployment (X_1)	Renter-occupied Housing (X_2)	Racial Segregation (X_3)	Young Male (X_4)
Unemployment (X_1)	1.000	0.175	0.068	−0.105
Renter-occupied Housing (X_2)	0.175	1.000	0.181	0.465
Racial Segregation (X_3)	0.068	0.181	1.000	0.183
Young Male (X_4)	−0.105	0.465	0.183	1.000

The highest correlation is 0.465, between renter-occupied housing units (X_2) and young male population (X_3). This does not seem so high as to suggest a collinearity problem, and none of the other correlations are at all remarkable.

She then ran the multivariate regression analysis with X_1, X_2, X_3, and X_4 as independent variables. The multiple regression results from these data are below (with "SE" written for "standard error"):

$R^2 = 0.509$

df for t = 19

	b	SE of b	t	p-value
Intercept (constant)	-392.009	218.221	-1.796	0.088
Unemployment (X_1)	37.081	22.891	1.620	0.122
Renter-occupied housing (X_2)	13.945	4.685	2.976	0.008
Racial segregation (X_3)	3.598	2.385	1.509	0.148
Young male population (X_4)	-23.818	10.334	-2.305	0.033

R^2 was only moderately high, with a value of 0.509. Such an R^2 generally indicates a reasonably satisfactory fit of the regression model to the data, but here does not seem so high in light of the small sample size (N = 24). 50.9% of the variance in violent crime rates (Y) in this sample was explained, or accounted for, by this set of X's (X_1 through X_4), and for the most part Y and $\hat{Y}$ were somewhat close for cities in the sample. Still, it appears that there are additional independent variables that should be considered in order to provide a more complete explanation for differences in crime rates among Colorado cities.

Checking for p-values below the usual 0.05 cutoff indicates that the effects of two of the four X's were statistically significant. Renter-occupied housing (X_2) and young male population (X_4) had statistically significant effects on violent crime, controlling for other X's. p-values for unemployment (X_1) and the racial segregation index (X_3) were above the 0.05 cutoff. Thus, when controlling for other X's, the effects of unemployment (X_1) and racial segregation index (X_3) on violent crime (Y) were not statistically significant.

The regression equation based on this output is

$$\hat{Y} = -392.009 + (37.081)X_1 + (13.945)X_2 + (3.598)X_3 + (-23.818)X_4$$

or

$$\hat{Y} = -392.009 + 37.081X_1 + 13.945X_2 + 3.598X_3 - 23.818X_4.$$

For a and the b's we have retained three decimal places from the output, but it would also be fine if we reported more rounded-off figures, especially if we wanted to avoid giving an undue impression of precision in our results. Here the three decimal places are helpful for readers who run this analysis in their software and want to confirm that they obtained exactly the same results; we do this throughout the book, but normally more rounding is fine when presenting results.

Because the effects of renter-occupied housing (X_2) and young male population (X_4) were statistically significant, the researcher proceeded to interpret the slopes (b's) for renter-occupied housing (b_2) and young male population (b_4). b_2 was 13.945, indicating a positive effect of renter-occupied housing (X_2) on violent crime rate (Y). The predicted violent crime rate (per 100,000 population) increases by 13.945 as renter-occupied housing (X_2) increases by 1% (because this particular X_2 is measured as a percentage, in this instance "one unit" of X_2 means 1%), controlling for the other X's in the model.

To examine whether the increase of 13.945 is large enough to have a meaningful impact in real terms (real-life significance), the researcher should first examine the descriptive statistics for renter-occupied housing units (X_2) and violent crime rates (Y) produced in her statistical software. Renter-occupied housing's (X_2) sample mean is 40.34, with a standard deviation of 7.29 and a range of 26.70 to 52.30. This context suggests that it may be more informative to look at something more than a one-unit increase in renter-occupied housing (X_2). This does not fundamentally change the regression results but rather allows us to make a more reasoned interpretation, because the descriptive statistics for renter-occupied housing (X_2), including the standard deviation and range, suggest that a one-unit increase is rather small. A five-unit increase seems like a more meaningful change in renter-occupied housing (X_2). To interpret with a five-unit increase, she needs to multiply b by five: a five-unit increase in renter-occupied housing (X_2) increases the predicted violent crime rate ($\hat{Y}$) by $(5 \times 13.945) = 69.725$. Next, in light of the mean (342.17) and the range (116.1–854.2) of violent crime rates (Y), an increase of 69.725 in predicted violent crime rate ($\hat{Y}$) for a 5% increase in the renter-occupied housing seems large enough to have a real impact. Therefore, she can say that the effect of renter-occupied housing (X_2) on violent crime rate (Y) has real-life significance. Although this is ultimately a judgment that the researcher is making, and can be challenged by other researchers, it is rooted in a close examination of the sample data.

Suppose that b_2 had been 1.395, rather than the actual figure of 13.945. Then a five-unit increase in renter-occupied housing would have implied an increase of 6.975 (from 5×1.395) in the predicted violent crime rate. Many analysts would view such a change as rather small in light of the mean (342.17) and range (116.1–854.2) of violent crime rates in the sample, and the researcher would likely then conclude that renter-occupied housing's effect did not have much real-life significance. More generally, we always should be mindful of the possibility that a statistically significant effect is not large enough to demonstrate real-life significance. As noted earlier, this is especially a concern in analyses of very large samples; because standard errors tend to be smaller (and, in turn, t-statistics larger and p-values smaller) in larger samples, some effects that are quite small, or even trivial, in real terms can be statistically significant.

b_4 was −23.818, indicating a negative effect of the young male population (X_4) on violent crime rate (Y). The predicted violent crime rate (per 100,000 population) decreases by 23.818 as young male population (X_4) increases by 1%, controlling for the other X's. Young male population's (X_4) mean is 8.26, and its range is 5.20 to 16.80, so a 1% increase in young male population does not seem so small as to be uninterpretable. Relative to the mean (342.17) and range (116.1 to 854.2) of violent crime rates (Y), a decrease of 23.818 in the violent crime rate seemed to the researcher to be large enough to say that the effect of young male population (X_4) on violent crime rate (Y) has real-life significance.

With these results, the researcher was able to evaluate the research hypothesis. Based on the earlier examination of p-values and slopes (b's), renter-occupied housing (X_2) is the only X that showed a statistically significant (and also real-life significant) effect on crime (Y) in society in the direction—violent crime rate increases as renter-occupied housing increases—that was expected under the research hypothesis. The effect of young male population (X_4) was statistically significant (and also real-life significant), but in the negative direction—violent crime rate decreases as young male population increases—which is opposite of what the research hypothesis suggested. R^2 is moderately high, indicating that the set of X's is explaining crime (Y) fairly well, but it would be worthwhile for further analyses to explore additional X's that might also influence crime (Y). Note, though, that there is a practical constraint on the number of X's here, because as a general rule one wants the number of X's to be relatively small compared to the sample size N. The small sample size here (N = 24) means that it will not be appropriate to include a very large number of X's in the regression.

Overall, then, the research hypothesis was not very well supported by the regression analysis results for these data. Of course it is possible that the measurements of variables, or level of analysis, are not ideal. For example, economic deprivation could instead be measured by the percentage of city residents living in poverty, or racial inequality could be measured as the difference between average incomes for white and non-white residents. It is also possible that a lower level of analysis, such as neighborhoods, would provide a better test of the hypothesis. In addition, the counterintuitive finding of a negative relationship between young male population (X_4) and violent crime rate (Y) may not have arisen had the analysis controlled for other important X's. Among these cities, Boulder, Ft. Collins, and Golden have exceptionally high young male populations because these cities are home to the University of Colorado, Colorado State University, and the Colorado School of Mines, respectively. None of these cities have especially high violent crime rates, so without controlling for other variables that could capture the distinctive "college town" character of these cities, the regression analysis takes the data for these cities as evidence for a negative relationship between young male population and violent crime rate.

The researcher found a positive and significant relationship between renter-occupied housing and crime from these city-level (aggregate-level) data. She can make only aggregate-level interpretations from these findings. That is, it is legitimate to say that the analysis indicates that, holding other X's constant, cities with a high proportion of renter-occupied housing units are predicted to have higher crime rates than cities with a low proportion of renter-occupied housing units. However, it is not legitimate to make an individual-level interpretation of these results. This city-level analysis does *not* show that renters are more likely to commit crime than home owners; that would be the ecological fallacy of drawing individual-level conclusions from the aggregate-level results.

The p-values in the table of regression results for this example are two-sided, as they refer to the alternative hypothesis $\beta \neq 0$; this usually is the case with the default output from statistical software. Note that the original research hypothesis specified positive signs for the relationships between independent variables and the dependent variable. The researcher could appeal to various theories of crime to argue that the negative relationships between these independent variables and crime would be very unlikely. Recalling the review of hypothesis tests in Chapter 1, one might then suggest using one-sided rather than two-sided p-values in this situation, with H_1 specifying the sign of the effect if the null hypothesis of $\beta = 0$ is rejected. These one-sided p-values can be obtained by simply dividing the reported two-sided p-values by 2, which of course makes the one-sided p-values smaller. Statistically significant relationships with the two-sided p-value ≤ 0.05 would remain so in the one-sided approach, but some previously nonsignificant relationships might become statistically significant.

In the analysis here, the one-sided p-values for the previously nonsignificant independent variables unemployment (X_1) and racial segregation (X_3) would remain over the 0.05 cutoff, but close enough (0.122 / 2 = 0.061 and 0.148 / 2 = 0.074, respectively) to perhaps be viewed as borderline or marginally significant. However, this example also illustrates our caution about using the one-sided p-value in many research settings. The seemingly impossible negative effect of young male population on crime was in fact observed in the researcher's analysis of these data, despite the strong belief among researchers that societies with large young male populations tend to have high levels of violent crime. As discussed earlier, we can understand why this counterintuitive result was found in this particular sample and how it might change in the presence of other controls. Still, this illustrates how we may not always be so confident that we can absolutely rule out either a positive or negative effect when applying our understanding of theory or past research to new situations or data. We therefore will reserve the one-sided approach for analyses in which we have an unusually strong basis for ruling out one sign as impossible for the true effect of X on Y, even if our initial expression of the research hypothesis suggests a particular direction of the effect.

Finally, these data were obviously not derived from a true experiment that could be manipulated by the researcher, but instead came from observation of the world as it is. Of course there would be no possible way for the researcher to randomly assign different levels of renter-occupied housing and the other independent variables to different cities. Therefore the effect of renter-occupied housing on crime found by the researcher here should be understood as indicating an association between these variables. This *could* reflect a causal effect of renter-occupied housing on crime, but the nature of the research design does not allow the researcher to conclude that definitively.

2.5.2 Example 2: Infant Mortality in the World

A researcher believed that greater economic and technological development, health care availability, and urbanization decrease infant mortality in the world's nations and wanted to test this hypothesis with a multiple regression analysis. Further, the researcher was interested in exploring this question via historical rather than contemporary data, as theoretical arguments suggested that these relationships could be different in different historical periods. For this analysis, independent variables were economic and technological development, health care availability, and urbanization, and the dependent variable was infant mortality. He decided to use 28 randomly selected countries as the cases, so this research was at the country level (aggregate level of analysis). Data were obtained from a reference book that collected data from various original sources and reflected conditions around 1990.

Economic development was measured by gross domestic product (GDP, essentially the value of goods and services produced in the country's economy) per capita (or per person) expressed in U.S. dollars. A high value of GDP per capita in a country indicates a high level of economic activity and development. Technological development was measured by the number of people per telephone in the country; note that these data were collected before the cell phone era. A high number of people per telephone in a country indicates a low level of technological development, because more people "sharing" each phone means that there are, relative to the population, fewer phones and less access to technology. Health care availability was measured by the number of people per hospital bed. As with the telephone measure, a high number of people per hospital bed in a country indicates a low level of health care availability, as this means that there are few hospital beds relative to the population. Urbanization was measured by the percentage of population living in urban areas. It hardly needs to be said that a country with a high percentage of population living in urban areas is highly urbanized. Finally, infant mortality was measured by the number of infant deaths per 1,000 births.

The research hypothesis, units of analysis, measurements of independent and dependent variables, and raw data are as follows:

- *Research hypothesis*: Greater economic and technological development, health care availability, and urbanization decrease infant mortality in society.
- *Units of analysis*: 28 countries (aggregate-level data).

Measurements of independent and dependent variables:

- X_1: GDP per capita in U.S. dollars (economic development).
- X_2: Number of people per telephone (technological development).
- X_3: Number of people per hospital bed (health care availability).
- X_4: Percentage of population living in urban areas (urbanization).
- Y: infant mortality rate per 1,000 births (infant mortality).

Research hypothesis with measurements of independent and dependent variables:

1. GDP per capita (X_1) has a negative effect on infant mortality rate (Y).
2. The number of people per telephone (X_2) has a positive effect on infant mortality rate (Y). Countries with more people "sharing" a phone (X_2), reflecting less access to technology, will have higher infant mortality rates (Y).

3. The number of people per hospital bed (X_3) has a positive effect on infant mortality rate (Y). Countries with more people per hospital bed (X_3), indicating less health care availability, will have higher infant mortality rates (Y).

4. The percentage living in urban areas (X_4) has a negative effect on infant mortality rate (Y).

Data:

Country ID	Country Name	Infant Mortality (Y)	GDP per Capita (X_1)	People per Telephone (X_2)	People per Hospital Bed (X_3)	Percentage Living in Urban Areas (X_4)
1	Angola	151	950	132	845	29
2	Bangladesh	112	200	572	3,195	24
3	Bolivia	83	690	37	685	51
4	Burkina Faso	117	205	482	1,359	8
5	China	33	360	89	428	27
6	Cyprus	10	7,585	2	165	69
7	Ecuador	60	1,070	28	610	54
8	Ethiopia	113	130	320	3,873	11
9	Germany	7	14,600	1.5	95	86
10	Guyana	51	300	47	341	35
11	Indonesia	70	630	172	1,485	31
12	Jamaica	17	1,400	13	468	52
13	Liberia	119	440	278	800	46
14	Madagascar	93	200	239	600	22
15	Mauritius	22	2,300	15	364	41
16	Morocco	56	1,060	62	959	50
17	Netherlands	7	16,600	1.6	164	88
18	Pakistan	105	380	131	1,706	32
19	Poland	14	4,300	7.5	154	62
20	Saudi Arabia	69	5,800	13	406	78
21	South Korea	23	6,300	3.3	429	74
22	Spain	6	12,400	2.5	198	79

(Continued)

Country ID	Country Name	Infant Mortality (Y)	GDP per Capita (X_1)	People per Telephone (X_2)	People per Hospital Bed (X_3)	Percentage Living in Urban Areas (X_4)
23	Syria	45	2,300	17	840	50
24	Turkey	54	3,400	7	465	61
25	United Kingdom	8	15,900	1.9	138	90
26	United States	10	22,470	1.9	198	76
27	Venezuela	23	2,590	11	370	83
28	Zambia	77	380	78	311	49

The researcher first ran a correlation analysis in his statistical software, to check correlations among the X's in order to detect potential collinearity problems. The results of the correlation analysis were:

	GDP per Capita (X_1)	People per Telephone (X_2)	People per Hospital Bed (X_3)	Percent Living in Urban Areas (X_4)
GDP per Capita (X_1)	1.000	−0.439	−0.431	0.755
People per Telephone (X_2)	−0.439	1.000	0.767	−0.728
People per Hospital Bed (X_3)	−0.431	0.767	1.000	−0.656
Percent Living in Urban Areas (X_4)	0.755	−0.728	−0.656	1.000

The correlation analysis indicated that percentage living in urban areas (X_4) is strongly correlated with GDP per capita (X_1) ($r = 0.755$), the number of people per telephone (X_2) ($r = -0.728$), and the number of people per hospital bed (X_3) ($r = -0.656$). The number of people per telephone (X_2) and the number of people per hospital bed (X_3) are also strongly correlated ($r = 0.767$). These high correlations suggest that collinearity may indeed be a problem in the regression analysis.

Results of the multiple regression with all X's included were as follows, with "SE" again written for "standard error":

$R^2 = 0.673$

df for t = 23

	b	SE of b	t	p-value
Intercept (constant)	67.718	24.807	2.730	0.012
GDP per capita (X_1)	−0.002	0.001	−1.556	0.133
People per telephone (X_2)	0.116	0.061	1.899	0.070
People per hospital bed (X_3)	0.004	0.009	0.386	0.703
Percent living in urban areas (X_4)	−0.333	0.444	−0.750	0.461

The results show the classic symptom of collinearity that we discussed earlier. The R^2 is fairly high (0.673), suggesting rather successful predictions of Y in the sample, but none of the effects of X on Y are statistically significant (although the p-value for X_2, number of people per telephone, is close to the 0.05 cutoff). The researcher then tried to address the collinearity problem by excluding an X from each of the highly correlated pairs and rerunning the regression analysis.

Because percentage living in urban areas (X_4) is strongly correlated with three other X's (GDP per capita [X_1], the number of people per telephone [X_2], and the number of people per hospital bed [X_3]), it seemed prudent to exclude percentage living in urban area (X_4) from the new regression model. With the number of people per telephone (X_2) and the number of people per hospital bed (X_3) also strongly correlated, it would be reasonable to exclude one of those two also. The researcher chose to exclude the number of people per hospital bed (X_3) along with percentage living in urban areas (X_4). His next regression model then included only GDP per capita (X_1) and the number of people per telephone (X_2).

This seems to restate the overall research hypothesis as "greater economic and technological development decrease infant mortality in society," and the specific hypotheses as "GDP per capita (X_1) has a negative effect on infant mortality rate (Y)," and "the number of people per telephone (X_2) has a positive effect on infant mortality rate (Y)." Note, however, that the choice of which X's to exclude in response to collinearity was rather arbitrary. With health care availability and urbanization measures excluded for this reason, the researcher must be careful with interpretation. If results show statistically significant effects of the economic and technological measures, that does not rule out health care availability and urbanization as factors affecting infant mortality. Rather collinearity is making it too hard to distinguish all these different effects and forcing the researcher to be a bit more modest in his research goals.

Regression results with GDP per capita (X_1) and the number of people per telephone (X_2) are as follows:

$R^2 = 0.661$

df for t = 25

	b	SE of b	t	p-value
Intercept (constant)	52.430	8.138	6.443	0.000 (<0.001)
GDP per capita (X_1)	−0.003	0.001	−3.098	0.005
People per telephone (X_2)	0.157	0.037	4.265	0.000 (<0.001)

R^2 is little changed, with 66.1% of the variance in infant mortality (Y) explained by GDP per capita (X_1) and the number of people per telephone (X_2). Excluding the number of people per hospital bed (X_3) and percentage living in urban areas (X_4) from the regression did not reduce R^2 very much. That is, the new regression model's fit to the sample data is only slightly worse than when all four X's were included.

With number of people per hospital bed (X_3) and percentage living in urban areas (X_4) removed, effects of GDP per capita (X_1) and the number of people per telephone (X_2) on the predicted infant mortality rate (Y) become statistically significant. The p-value for GDP per capita (X_1) is 0.005, and the p-value for the number of people per telephone (X_2) is shown by some statistical software as 0.000, which can be interpreted as < 0.001 (it is not literally zero, but more decimal places would be needed to show that). Both are well below the 0.05 cutoff.

With this statistical significance, we can interpret effects (slopes) for GDP per capita (X_1) and the number of people per telephone (X_2). For GDP per capita (X_1), the slope b_1 is −0.003. This indicates a negative effect of GDP per capita (X_1) on the predicted infant mortality rate (Y) and that the predicted infant mortality rate per 1,000 births (Y) decreases by 0.003 as GDP per capita (X_1) increases by $1, controlling for the number of people per telephone (X_2). Of course $1 represents a very small change in GDP per capita (X_1), given the mean ($4,462.14) and range ($130–$22,470) of this variable's values in this data set. A more helpful interpretation might be that the predicted infant mortality rate per 1,000 births (Y) decreases by 3 as GDP per capita (X_1) increases by $1,000. The value 3 was obtained by multiplying b_1 by the $1,000 change in X_1: if a $1 increase in GDP per capita decreases the predicted infant mortality by 0.003, then a $1,000 increase in GDP per capita will decrease the predicted infant mortality by 1,000 × 0.003, or 3. Put this way, the effect does not seem so tiny, considering the mean (55.54) and the range (6–151) of infant mortality rate (Y) in the sample, even if it is still not very large. Thus, the effect of GDP per capita (X_1) appears to be meaningful in real-world terms.

For number of people per telephone (X_2), the slope (b_2) is 0.157. This is a positive effect of the number of people per telephone (X_2) on the predicted infant mortality rate (Y), and the predicted infant mortality rate (Y) increases 0.157 per 1,000 births as the number of people per telephone (X_2) increases by one, controlling for GDP per capita (X_1). Again it is interesting to consider a somewhat larger change in the number of people per telephone (X_2), as there is a fairly large mean (98.76) and very wide range of values (1.5–572) of X_2 in the sample. If the number of people per telephone (X_2) increases by 10, the predicted infant mortality rate (Y) increases by 1.57 (calculated as 10 × 0.157) per 1,000 births, still controlling for GDP per capita (X_1). As with the effect of GDP per capita, this change now does not seem so tiny relative to the sample mean (55.54) and range (6–151) of infant mortality rate (Y) and so might be assessed as having real-world significance. However, some researchers might look at that change as still being too small to indicate real-world significance. As discussed earlier, there can be ambiguity in the assessment of real-life significance.

The researcher then evaluated the new research hypotheses. With R^2 fairly high, the model does pretty well at predicting levels of infant mortality for the countries in the sample. R^2 is not so close to 1 as to suggest that all important variables have been included, but the model is reasonably successful. Further, both GDP per capita (X_1) and number of people per telephone (X_2) have statistically significant effects on the infant mortality rate (Y), and both effects have the expected signs: negative for GDP per capita, and positive for number of people per telephone. Both effects are also judged to be large enough to suggest real-life significance, though with some ambiguity for number of people per telephone. The revised research hypotheses thus seem to be supported in this analysis. Still, it is important to keep in mind the collinearity-induced choices of X's from the original set and realize that similar results likely would have been found if different choices of which independent variables to keep and drop had been made.

The impact of different choices of which X's to keep in the face of collinearity can be illustrated by considering these data further. Remember that the number of people per telephone (X_2) and the number of people per hospital bed (X_3) were strongly correlated ($r = 0.767$), and the researcher chose number of people per hospital bed (X_3) to exclude from the regression model. What if the other choice—excluding the number of people per telephone (X_2)—had been made? Here is the output from the regression model with independent variables GDP per capita (X_1) and number of people per hospital bed (X_3):

$R^2 = 0.572$

df for t = 25

	b	SE of b	t	p-value
Intercept (constant)	53.224	10.043	5.300	0.000 (<0.001)
GDP per capita (X_1)	−0.003	0.001	−3.134	0.004
People per hospital bed (X_3)	0.021	0.007	3.027	0.006

R^2 is somewhat smaller than in the previous regression but still moderately high. And, as in the previous model, both X's show statistically significant effects. We can interpret the slopes for X_1 and X_3. Predicted infant mortality rate per 1,000 births (Y) decreases by 0.003 as GDP per capita (X_1) increases by $1, controlling for the number of people per hospital bed (X_3). (As before, a more helpful interpretation for assessing real-life significance might focus on an increase of $1,000 in GDP per capita.) This is the same effect of GDP per capita (X_1) as was estimated in the previous regression.

For number of people per hospital bed (X_3), the slope b_3 is 0.021. That is, predicted infant mortality per 1,000 births increases by 0.021 as the number of people per hospital bed (X_3) increases by one, controlling for GDP per capita (X_1). Again it is interesting to consider a somewhat larger change in the number of people per hospital bed (X_3), as there is a large mean (773.25) and very wide range of values of X_3 (95–3,873) in the sample. For an increase of 100 in the number of people per hospital bed (X_3), the predicted infant mortality rate (Y) increases by 2.1 per 1,000 births, while holding GDP per capita (X_1) constant. When considered this way, a reasonable change in the number of people per hospital bed produces a meaningful change in predicted infant mortality, suggesting real-world significance of this effect. An increase in the number of people per hospital bed indicates a decrease in the availability of health care, so, controlling for economic development, predicted infant mortality increases as health care availability decreases. This is consistent with the original hypothesis about infant mortality and health care availability.

The regression results with GDP per capita (X_1) and number of people per telephone (X_2) are quite similar to those with GDP per capita (X_1) and the number of people per hospital bed (X_3). Although R^2 is somewhat higher in the regression using number of people per telephone (X_2), in each regression both independent variables have statistically significant effects. Also, the size and direction of the effect of GDP per capita (X_1) are the same in the two regressions, and whether number of people per telephone (X_2) or number of people per hospital bed (X_3) is used, the effect of this variable is statistically significant and positive.

This similarity in results is not surprising. Because highly correlated variables represent similar information, using one or the other in the regression will often produce similar results. This is why we should be cautious in discussing results for regressions carried out after removing variables due to collinearity. We should not give the impression that one or the other of these regressions definitively establishes that either technological development (represented by X_2) or health care availability (represented by X_3) is the factor that is truly related to infant mortality (Y). Instead, the collinearity limits us to deciding that one or both of these factors appear to be related to infant mortality, but the nature of the data prevents a more conclusive interpretation. (We will revisit this idea as one application of the F-statistic in Chapter 3.) And as before, the nonexperimental research design does not permit us to ascribe a definitive causal interpretation to these effects.

2.6 Exercises

2.6.1 Exercise 1: Median Income Among U.S. States

A researcher hypothesized that higher unemployment and a larger older adult population decrease the typical income in a given area. On the other hand, she hypothesized that greater educational attainment and urbanization increase an area's typical income. To test these research hypotheses, she collected data from 34 American states.

Typical income was measured as the state's median annual income in dollars, and unemployment was measured as the percentage of its labor force that is unemployed. Older adult population was measured as the percentage of population aged 65 years and over, with educational attainment measured as the percentage of those 25 years and older holding at least a bachelor's degree. Finally, urbanization was measured as population density (population per square mile). The following data were collected from the Census Bureau:

ID	State	Median Income	Unemployment	Older Adult Population	Educational Attainment	Urbanization
1	Alabama	48,123	5.8	16.5	25.5	96.3
2	Arizona	56,581	5.8	17.1	29.4	61.8
3	Arkansas	45,869	5.6	16.5	23.4	57.7
4	California	71,805	5.9	13.9	33.6	253.8
5	Colorado	69,117	4.2	13.8	41.2	54.1
6	Connecticut	74,168	6.1	16.8	38.7	741.1
7	Delaware	62,852	5.3	18.0	31.5	493.6
8	Florida	52,594	5.5	20.1	29.7	391.3
9	Georgia	56,183	5.8	13.4	30.9	181.3
10	Illinois	62,992	6.1	15.2	34.4	230.6
11	Indiana	54,181	4.7	15.4	26.8	186.1
12	Iowa	58,570	3.6	16.7	28.9	56.3
13	Kentucky	48,375	5.5	15.9	24.0	112.8
14	Louisiana	46,145	6.5	14.9	23.8	108.4
15	Maryland	80,776	5.2	14.9	39.7	623.5
16	Massachusetts	77,385	4.6	16.1	43.4	879.5
17	Michigan	54,909	5.9	16.7	29.1	176.2
18	Minnesota	68,388	3.6	15.4	36.1	70.0
19	Mississippi	43,529	7.0	15.6	21.9	63.6
20	Missouri	53,578	4.6	16.5	29.1	88.9
21	New Hampshire	73,381	3.8	17.6	36.9	150.0
22	New Jersey	80,088	5.3	15.7	39.7	1,224.6
23	New York	64,894	5.5	15.9	36.0	421.2
24	North Carolina	52,752	5.3	15.9	31.3	211.3

(*Continued*)

ID	State	Median Income	Unemployment	Older Adult Population	Educational Attainment	Urbanization
25	Ohio	54,021	5.2	16.6	28.0	285.3
26	Oklahoma	50,051	5.4	15.3	25.5	57.3
27	Pennsylvania	59,195	5.3	17.8	31.4	286.2
28	Rhode Island	63,870	5.7	16.7	33.5	1,024.8
29	South Carolina	50,570	5.8	17.2	28.0	167.1
30	Tennessee	51,340	4.9	15.9	27.3	162.9
31	Texas	59,206	5.1	12.2	29.6	108.4
32	Vermont	57,513	3.8	18.8	38.3	67.7
33	Virginia	71,535	4.6	15.0	38.7	214.5
34	Washington	70,979	4.9	15.1	35.5	111.4

Year of data: 2017
Data source: Census Bureau

1. What are the independent and dependent variables for this analysis?
2. After entering the above data in statistical software, check for collinearity and then run multiple regression analysis as needed to test the research hypothesis.
3. Interpret the results (output) as fully as possible, focusing on R^2, p-values, b's, and real-life significance (remember that descriptive statistics are helpful in making interpretations of real-life significance).
4. Evaluate the research hypothesis.
5. Use the results to predict median income for a state with 4.3% unemployment, 18.1% older adult population, 35.5% holding a bachelor's degree, and 260.3 people per square mile. Should you trust this predicted value? Explain.

2.6.2 Exercise 2: Predicting Educational Attainment

Researchers were interested in studying influences on educational attainment, and obtained data on a sample of American adults who had completed their formal education. They believed that an adult's education could be predicted from measurements taken at childhood on his or her reading test score, resource competition in the family, parental education, and household income. In the following hypothetical descriptive and regression analyses of the data from this sample, the dependent variable is years of education, and the independent variables are reading test score, number of siblings (measuring resource competition), parental education (whichever parent had the highest, in years), and household annual income for the respondent at age 15 (in dollars).

Descriptive Statistics:

	Mean	Minimum	Maximum
Respondent's education	14.3	9	21
Test score	100	82	128
Number of siblings	2.6	0	6
Parental education	12.2	8	18
Childhood household income	54,567.12	24,000	78,000

Multiple Regression Analysis Results:

$R^2 = 0.228$

df for t = 103

	b	SE of b	t	p-value
Intercept (constant)	−95.5			
Test score	0.028	0.017	(i)	0.103
Number of siblings	−0.121	0.058	(ii)	0.039
Parental education	1.122	0.580	(iii)	0.056
Childhood household income	0.002	0.001	2.000	0.048

1. Assuming no collinearity, interpret the results as fully as possible, focusing on R^2, p-values, b's, and statistical and real-life significance.
2. What are the values for the t-statistics (i), (ii), and (iii) not shown in the table?
3. What are the predicted years of education for a person with a test score of 105, three siblings, whose most-educated parent had 12 years of education, and whose household income was $46,000 when they were 15 years old? Should you trust the predicted value?
4. What is the total number of cases in the sample?
5. Is the research hypothesis supported? Explain.

2.6.3 Exercise 3

1. Suppose that our research hypothesis is represented by the set of independent variables in the regression. Explain why both R^2 and statistical significance of the b's are important in deciding whether the research hypothesis is supported.

2. If we believe that there are multiple factors that influence our dependent variable, why do we need to use multiple regression instead of repeatedly applying simple (bivariate, with just one independent variable) regression?

3. Explain how collinearity can lead to artificial findings of statistical nonsignificance of X's.

2.7 Appendix A: Beginning a Research Project Using Multiple Regression

2.7.1 Setting Up the Research Question and Hypothesis

This chapter presented core interpretations of multiple regression results. Here we step back and consider how to start a research project using multiple regression analysis. We begin with a research question and a research hypothesis about the expected influences of the multiple independent variables on the dependent variable. Social science theories often suggest such relationships, and so theories can help guide us in choosing the independent and dependent variables. Additional independent variables may be suggested by previous quantitative research, or even just common sense. Note that we are using "research hypothesis" to mean an overall statement of expected influences on the dependent variable. This is different from the specific statistical hypotheses (H_0 and H_1) that we discussed in the main text of this chapter. Of course specific statistical hypothesis tests will be helpful in evaluating the research hypothesis.

When a theory is guiding the analysis, independent variables that do not have a central role in the theory are sometimes called *"control" variables*. This distinguishes them from the variables that are the main elements of the theory and the focus of the research hypothesis. The researcher expects them to be related to Y and likely correlated with at least some of the independent variables that are the theory's main focus, and so wants to control for them when determining the effects of the independent variables of primary interest. Despite this conceptual distinction, however, the control variables are still treated the same as other independent variables in the regression analysis itself, and the control variables play the same part in predicting Y as do the other independent variables. It is just that results for the control variables may not be emphasized as much when writing a report or article on the research.

For example, in criminology Shaw and McKay's (1942) social disorganization theory suggests that greater poverty, residential instability, and cultural conflict increase crime in a community. In this case, the dependent variable is crime, and the independent variables suggested by the theory will measure poverty, residential instability, and cultural conflict. In addition, much previous research has found that areas with more young males—an especially crime-prone group—tend to have higher crime rates, and the size of the young male population may also be correlated with the independent variables suggested by the theory. Thus young male population should be included in the regression analysis as another independent variable, but with respect to the theory it is a control variable. In this case, then, the research hypothesis would be "The levels of poverty, residential instability, cultural conflict, and young male population in a community all have positive effects on community crime rates." Here the research hypothesis specified the direction of the relationships (in this case, positive) between the independent variables and the dependent variable, but sometimes the research hypothesis only says that there is a relationship, leaving the direction unspecified.

2.7.2 Level of Analysis

Theories can also suggest whether an aggregate-level or an individual-level analysis is most appropriate for our research. As discussed in Chapter 1, level of analysis refers to the nature of the subjects or units we are studying. In the social disorganization example above, the theory argues that communities with high levels of poverty, residential instability, and cultural conflict are more likely to have high crime rates. This suggests the community as the appropriate unit to study, yielding an aggregate-level analysis. The theory does not necessarily imply that, within a community, crime is being committed by individuals who are poor or have recently moved; an individual-level theory of crime would be needed for that sort of investigation. (In fact theories sometimes encompass more than one level; we briefly discuss multilevel data in Chapter 9.)

In this way, social disorganization theory is tied to aggregate-level analysis, using data from communities such as neighborhoods, cities, counties, or states. An individual-level theory such as Hirschi's (1969) social control theory, on the other hand, should be examined with data on individuals. As we noted in Chapter 1, in practice it is usually the case that data from higher levels, such as states, are easier to obtain than data from lower levels, such as neighborhoods or individuals. This makes it especially important to be careful not to assume that results obtained from data at one level of analysis automatically apply to another level. Recall the "ecological fallacy" from Chapter 1, in which findings from aggregate-level research are inappropriately applied to individual-level phenomena. Because aggregate-level data are often more accessible, it can be very tempting to draw individual-level conclusions from aggregate-level data, but this temptation should be resisted. It is important to use data that match the level of analysis suggested by the theory we are investigating and to interpret results in terms of that level.

2.7.3 Measuring Independent and Dependent Variables

Our next step is to think how to actually measure the independent and dependent variables for the subjects or units we are studying. Again we can take advantage of previous quantitative research and common sense in coming up with measurements of the variables. For the social disorganization theory example, poverty is often measured by the percentage of households in a community whose total income falls below an official poverty line. Residential instability can be measured as the percentage of residents living in different housing than they were 5 years ago, or by the community's percentage of renter-occupied housing units. Cultural conflict is often measured by the community's percentage of foreign-born population. Young male population is simply the percentage of the population that is made up of men in a particular age range. Crime can be measured as the total crimes (or a more specific crime type such as homicides or robberies) reported to police in a community, converted into a rate per 100,000 population so that values from communities of varying size are comparable.

In any case, we want measurements that closely match the theoretical concept for each variable while still being practical to obtain for all the subjects or units in our study. When there is no available variable that can directly measure a particular theoretical concept, or there is no practical way to carry out that measurement in our sample, we must either find the best available alternative or not include a variable measuring that concept at all. In either case, this will be a limitation of our research and will be important to note in any papers or reports describing our work.

2.7.4 Data Collection

After deciding on the level of analysis, and appropriate measurements of the independent variables and the dependent variable, we can collect data on these measurements for the subjects or units of analysis in our sample. For the social disorganization example, suppose that we have decided to use states as our units of analysis; remember that the theory requires aggregate-level analysis of some kind. We can then collect data from all 50 American states on the percentage of households below the poverty line (the poverty measure) the percentage of residents living in different housing than 5 years ago or the percentage of renter-occupied housing units (the residential instability measure), the foreign-born population percentage (the cultural conflict measure), the population percentage of young men aged 15 to 21 years (the young male population measure), and the total crimes reported to police per 100,000 population (the crime measure). For the United States, a great variety of national or community (aggregate-level) data can be found at the websites of various government agencies, including, among others, the Census Bureau and the Bureau of Justice Statistics. An excellent source for individual-level data that can be used to investigate a wide variety of research questions is the University of Michigan's Inter-University Consortium for Political and Social Research (ICPSR) data archive.

After data collection, we enter data in our statistical software, and use the software to run the multiple regression analysis that we have decided will best address our research question and hypothesis. The software output will give us the numerical results described above and allow us to make interpretations. In some cases, there will be an intermediate step in which some variables need to be transformed before being used in the regression analysis, but we will discuss that situation in subsequent chapters.

2.7.5 Evaluation of Research Hypothesis

We want to use the regression results to evaluate the research hypothesis that we developed from our research question. How well is the research hypothesis supported in our data? To evaluate this, we go through the different elements of interpretation that we discussed in the preceding sections of this chapter. Recapping those sections, we need to consider (a) the value of R^2, (b) statistical significance of the effects of the X's (via the p-values), and (c) for those that are statistically significant, the slopes, or effects, of the X's (the b's) in terms of direction of the effects and their real-life significance. In general, we view the analysis as strongly supporting our research hypothesis if R^2 is high and all (or most of) the X's have statistically significant effects on Y that are in the expected (by the research hypothesis) direction, and, further, these effects are large enough to have real-life significance.

A high R^2 means that the set of X's suggested by the research hypothesis is predicting Y well and that we seem to be accounting for the main influences on Y. Statistical significance of every X's effect means that each of the hypothesized X's does appear to be related to Y, because for each X we are rejecting the statistical null hypothesis of no effect (H_0). For each X with a statistically significant effect, we can check the effect's direction (positive or negative) from the slope b and see if it matches the direction suggested by the research hypothesis (though again sometimes that direction is not specified by the research hypothesis). We can also determine from each b if the magnitude of the effect is enough to believe that the corresponding X has a real impact on Y (still realizing that a genuinely causal interpretation of this effect usually cannot be made).

Even if most, or all, of the X's have statistically significant effects on Y in the direction suggested by the research hypothesis, there could still be a low value of R^2. This typically means that the regression model is incomplete, in the sense that our regression is missing some other X's that are important in predicting Y, so that the research hypothesis is incomplete as an explanation for Y. In that case we may want to seek other X's based on theory, previous research, or common sense that could also affect Y and rerun the regression with these additional X's to see if R^2 improves. (Note too that a low R^2 could be due, at least in part, to the presence of nonlinear relationships between some X's and Y. We discuss such relationships in later chapters.) The opposite situation, with a high R^2 but none or almost none of the effects statistically significant, often indicates the problem of collinearity, which is discussed in this chapter.

The situation of a low R^2 along with none or almost none of the effects of the X's being statistically significant is quite damaging for the research hypothesis. In that case, the research hypothesis seems to be both missing important influences on Y **and** incorrect in suggesting that the X's it named are actually related to Y. This circumstance may be unlikely when we are deriving the research hypothesis from a well-developed theory, but sometimes even a well-developed theory is not supported when confronted with real data.

Along with discussing results, a paper reporting on a quantitative research project should also compare the current findings to those of previous research, discuss limitations, and, if applicable, suggest policy implications. In many cases it is also helpful to indicate future research directions that could build on the work. In the following section, we discuss a typical framework for organizing the parts of a quantitative research paper.

2.7.6 Organization of Quantitative Research Paper Using Multiple Regression

To help in writing a complete quantitative research paper, we list each part of such a paper below, following a standard framework that is typical of published research papers in the social sciences: (1) abstract, (2) introduction, (3) literature review, (4) research question, (5) data and measurement, (6) analysis and results of analysis, (7) conclusion, (8) tables and figures, and (9) references. Each part is discussed below. Note that the organization of the paper is separate from the specific format in which it is written. By "format" we mean details of how to present elements such as sections, section headings, citations, references, tables, and so on. Commonly used styles include those of the American Psychological Association (APA), American Sociological Association (ASA), and the Chicago Manual. Note that the organization of the paper laid out here is not universal in the social sciences, and examples of many variations on this structure can be found in the published literature.

1. **Abstract**
    - A short overall description of the paper, including results.
2. **Introduction**
    - Present your research topic (focusing on the relationship between X[s] and Y).
    - Be sure to mention the level of analysis for your research question (individual or aggregate).
    - Discuss why it is important to study this relationship (or relationships).
3. **Literature Review (Review of Theory and Previous Research)**
    - In social science, researchers' interest in the relationship between X(s) and Y is typically inspired by theories and/or previous research findings on the same, or closely related, topics.
    - Summarize relevant theories as well as findings of previous quantitative research. When summarizing previous research, focus on the research questions, the units being studied, the data sources, the types of data analysis, the independent and dependent variables, and the results/findings.
4. **Research Question**
    - Discuss your research question (topic) and research hypothesis (expectations for findings). It is often helpful to note the main differences between previous research and what you are doing, which could include points related to the data (such as the data source or when the data were collected) or the analytic methods being applied. When the goal is to eventually publish the paper

in a scholarly journal, the "novelty" of the research is a key part of the evaluation of the paper's potential importance. In other settings, novelty may not be emphasized so much. Also, sometimes a researcher working on a new topic or question for which little research or theory exists will not really have expectations for how results will look and so cannot state a research hypothesis. Exploratory work, therefore, will not always have this element.

5. **Data and Measurement**

    - Describe your data in detail, focusing on when it was originally collected, units (such as persons, cities, states, or nations), and data sources.
    - List independent (including control) and dependent variables and discuss how they were measured. For example: "Economic hardship is measured by the unemployment rate, obtained by dividing the number of unemployed people by the total number of people in the civilian labor force, times 100. The level of crime is measured by the number of reported homicides per 100,000 population."

6. **Analysis and Results of Analysis**

    - Use statistical software to run appropriate descriptive statistics for the independent and dependent variables involved in the multiple regression analysis and briefly discuss the descriptive statistics.
    - Diagnose collinearity and, if present, try to address it.
    - Discuss the statistical techniques being used.
    - Use statistical software to run multiple regression analysis; interpret and discuss results of this analysis, focusing on R-squared, p-values, and b's (including real-life significance).

7. **Conclusion (Implications of Analysis for Research Question and Hypothesis)**

    - Discuss whether your findings support the theoretical perspectives that you presented earlier in the paper and the research hypothesis posed earlier.
    - Compare your results to the findings of previous research. If there are differences that appear to stem from differences in data sources or analytic methods, it can be useful to discuss those, but this is not always the case.
    - If applicable, discuss policy implications.
    - Discuss the limitations of your research and future research possibilities.

8. **Tables and Figures**

    - Following the specific style required by a course instructor or the applicable style manual, present detailed results of analysis (descriptive and multivariate) in tables.

9. **References**

    - Follow the specific format required by a course instructor or the applicable style manual.

2.8 Appendix B: Additional Issues in Multiple Regression

2.8.1 Standardized Coefficients

This chapter has focused on the usual unstandardized coefficients, interpreted in terms of the units of measurement for the X's and Y. That is, if X_1 is years of education, and Y is dollars of income, b_1 is interpreted as the change, in dollars, in predicted income for an additional year of education, or when education increases by 1 year (holding other independent variables constant). This is very natural, because we have an intuitive grasp of the meaning of a certain change in dollars of income and the magnitude of the change represented by an additional year of education.

However, the fact that the X's in the multiple regression are typically measured in various different units makes it hard to compare the effects of different X's on Y. Continuing the example, suppose that X_2 is father's income. Certainly in any realistic sample the variance in people's incomes (measured in dollars) will be much greater than the variance in people's years of education (measured in years), so "1 more year of education" and "1 more dollar of father's income" are not comparable in any obvious way. Therefore the corresponding effects on Y are difficult to compare, and we cannot easily say which effect is more "important" in predicting Y. Even when we make a careful assessment of each independent variable's real-life significance, it may not be very clear which of the real-life significant effects are the largest.

The standardized coefficient is one response to this problem. This coefficient is a transformation of the usual b, so that the interpretation of an effect is made with respect to standard deviations of X and Y, not the original measurement units. That is, the standardized coefficient for the effect of education would indicate how many standard deviations—not dollars—predicted income would increase for each additional standard deviation—not year—of education. Standardized coefficients seem to allow for better assessment of which X's have the biggest effects on Y because the shift to standard deviations means that the interpretations no longer involve all the different and incomparable units in which the different X's are measured. This sounds as if it also would enhance assessments of real-life significance. An important difficulty, though, is that it is hard for us to think in units of standard deviations. Even if it is advantageous to put every variable's effect into a common framework, real-world significance still will often seem easier to assess in terms of the original units than in terms of standard deviations. We have a better grasp of the meaning of a change that is reported in the original units than of a change that is reported in units of standard deviations. The common framework of the standardized coefficients may assist with comparison of the effects' size but at the possible cost of less understanding of the effects' meaning.

There are also some technical objections to standardized coefficients, such as the possibility of the same b leading to very different standardized coefficients in different samples, due to differences across samples in the variables' variances. (More advanced texts discuss these technical points.) For these reasons, we will use unstandardized coefficients throughout the text. Still, it is valuable to understand what standardized coefficients are and how they are interpreted, as some researchers prefer to present multiple regression results in that form. Some articles and reports will show both the unstandardized and standardized coefficients in tables of regression results. The symbol β (or the word "beta") will sometimes be used to represent the standardized coefficient, but it is important to be clear that this is a different use than we are making of that symbol. As discussed above, we use β to symbolize a coefficient in the "true" regression model for the entire population, for which b is our estimate from the sample.

2.8.2 More on Diagnosing and Addressing Collinearity

Many statistical software packages include some more formal methods for diagnosing collinearity than simply examining correlations among the X's as we discussed above. For example, variance inflation factor (VIF) scores are an attempt to assess the extent to which collinearity is affecting standard errors of the regression coefficients by considering how strongly each X is related to the set of all other X's in the regression model. When high VIF scores are observed, collinearity is likely a problem, and there are various cutoffs in use for determining what is a high enough VIF score to indicate this.

When collinearity has been detected, one response is to create new variables that combine several highly correlated X's, using techniques such as *principal components* analysis or factor analysis. We discuss principal components a bit more in Chapter 9, but for now we just point out that the nature of the correlations among X's determines the construction of the new variables. Typically these new variables will, by virtue of the method used to construct them, be uncorrelated with each other, while still aiming to convey the information contained in the original X's. The new variables can then be used instead of the original X's in the regression, and by definition there will be no collinearity among independent variables in this new regression. The tradeoff for the desirable absence of collinearity when using these new variables is that b's for these constructed variables will be harder to interpret. The b's in the regression results will no longer refer to natural variables whose measurement and meaning feel intuitive to us, but instead to the constructed variables that have standardized scales. Advanced texts can be consulted for more guidance on this and other methods for detecting and addressing collinearity.

3

Categorical Independent Variables in Multiple Regression

Dummy Variables

In this chapter we introduce one of the most important extensions of the multiple regression framework discussed in Chapter 2. Many of the variables of greatest interest to social scientists are inherently *categorical* rather than numerical in nature. The regression framework presented in Chapter 2 can accommodate such variables, but some modification is needed. We first describe categorical variables, then discuss how categorical independent variables are represented in multiple regression via *dummy variables* and how this affects our interpretation of b for these variables and the assessment of statistical significance. Then we demonstrate multiple regression with categorical independent variables with a full example and conclude the chapter with exercises. Note that for now we are focusing on the use of categorical independent variables; categorical dependent variables will be introduced in Chapter 7.

3.1 Categorical Variables

Throughout the previous chapter, each X that we considered was numerical, representing some sort of quantitative measurement for which a numerical value was recorded for each unit in the sample. A categorical variable is different, in that the values indicate which type or category a case belongs in, rather than reporting a numerical measurement. It is easy to imagine that such variables play an important role in social science research: examples of common categorical variables include race (perhaps with categories African American, Asian, Native American, white, and other), ethnicity (with categories Latino and non-Latino), gender (in most existing data sets, categories male and female), religion (perhaps Catholic, Protestant, Muslim, Buddhist, and other), social class (such as high, middle, and low), political affiliation (often Democrat, Republican, and Independent), and region (for instance, North, South, East, and West).

It is important that the listed categories be *mutually exclusive* (no case falls into more than one category) and *exhaustive* (every case can be identified as belonging to a listed category). That sounds straightforward, but still for some variables it can be quite difficult in practice to decide exactly what categorization to use. Is the above categorization of social class into three categories (high, middle, and low) too crude? Would more categories better represent the concept of social class? For race, should "other" be replaced by additional specific categories, including those representing multiracial identities? Although these questions seem to suggest creating more categories, we also want to avoid creating categories that have very few members in our sample. Also, for variables like social class, increasing the number of categories could make the task of classifying people much more difficult. So some considerations may push us toward creating fewer (and broader) categories.

Another complication is that some concepts can be expressed through either a categorical or numerical variable. For example, in Chapter 2 education was measured quantitatively, as years of education. A categorical education variable, on the other hand, could indicate educational attainment in terms of diplomas and degrees (less than high school, high school, bachelor's degree, master's degree, doctorate). Likert scale responses from survey questionnaires (for example, strongly agree, agree, neutral, disagree, strongly disagree) are sometimes treated as numerical variables (for instance, with values 1, 2, 3, 4, 5), but in other instances it will seem preferable to consider them categorical. In the end, these kinds of decisions around categorization will need to take account of both the nature of our research goals and the characteristics of the sample we have.

Theory, previous research, or simple common sense will often suggest that categorical independent variables be included among the influences on the dependent variable in a multiple regression model. In an individual-level study of income, for example, educational degree, gender, and race/ethnicity are likely predictors of a person's income. As an aggregate-level example, subculture of violence theory notes regional differences in crime rates, with Southern states more apt to have high crime rates than non-Southern states. This would suggest including region as a categorical independent variable when analyzing state crime rates.

Despite the theoretical importance of including categorical independent variables in a regression, it is not immediately obvious how we would do that in practice. The framework developed in Chapter 2 expects numbers to be input as the values of the variables, not words such as "South." We might think that the solution is to create a coding scheme that identifies each region with a number, such as 1 for South, 2 for North, 3 for West, and 4 for East, and then simply treat region as a numerical independent variable in the regression. Such a scheme, however, introduces assumptions that may not be reasonable. First, it assumes that the predicted values of Y for members of the different categories, holding other independent variables constant, follow the order (or reverse order) in which we have assigned these values. In this example, a positive b for this variable would imply that the predicted values of Y are lowest for South, then higher for North, still higher for West, and highest for East. But of course the assignment of values was arbitrary; there is no reason why we could not have started out with South as 3 and West as 1, or any other variation on the coding scheme above. Because the order of categories implied by this coding scheme is arbitrary, we do not want to force the predicted values of Y to follow it.

Of course in some cases the categorical variable really does seem to be inherently ordered, such as for educational degree. For that variable, it seems more natural to order the categories from less to more educational attainment and then assign numerical codes that reflect this order. For example, we might say less than high school = 1, high school = 2, bachelor's degree = 3, and master's degree = 4. But even if we believe that the predicted Y for members of the different categories should follow this order, another questionable assumption is being made when we code the ordered categories with numerical values like 1, 2, 3, and so on. This coding assumes that the difference in $\hat{Y}$ between the category numbered 1 and that numbered 2 is the same as the difference between category 2 and 3, and so on; equal steps between the numbers representing the different categories imply equal steps in $\hat{Y}$ values for members of these categories. For educational attainment, one could imagine there being a large difference in $\hat{Y}$ (perhaps representing predicted income) between people with less than high school and those with high school, or between those with high school and those with bachelor's degrees, but only a small difference between those with bachelor's and those with master's degrees. So even when the categories have an inherent order, and we believe that this order should be reflected in the $\hat{Y}$ values, it may still be incorrect to further assume that the categories are equally spaced in terms of predicted Y.

To avoid these inappropriate assumptions, we prefer not to apply any coding scheme like this to categorical independent variables. Instead, we will make a special transformation of the categorical information into a numerical form, via dummy variables. Again remember that throughout this chapter we are examining categorical independent variables, not a categorical dependent variable (which will be taken up in Chapter 7).

3.2 Transforming a Categorical Independent Variable Into Dummy Variable(s)

Our transformation of a categorical independent variable involves construction of a special set of numerical variables, called *dummy variables*. The first step is to choose a reference, or left out, category. It does not matter which we choose as the *reference category*, and regression results will be fundamentally the same no matter which category is the reference. In particular, the choice of the reference category has no impact on R^2, results (meaning statistical significance and b) for the other independent variables in the regression, the order of highest to lowest $\hat{Y}$ (holding other variables constant) for the categories of the independent variable, and differences (again holding other variables constant) in $\hat{Y}$ between these categories. Though values of the b's associated with the dummy variables, and the intercept a, will change when a different reference category is used, this is purely a superficial difference in the regression output, not a true change in the results. Likewise, a categorical independent variable's statistical significance does not change when using a different reference category, but again there will be cosmetic differences in the output's p-values, as we will see in examples below. Still, even though these features of the regression results do not depend on the choice of reference category, it is important to remember which category we selected, because that will affect how we interpret and discuss the results.

After choosing a reference, we create a dummy variable for each of the remaining categories. This means that when transforming a categorical independent variable into a set of dummies, we will always have one less dummy than the number of categories; that is, the number of dummy variables is the total number of categories in the variable minus one. Each dummy variable has a simple form: it takes value "1" for subjects, cases, or observations that fall into the category represented by the dummy, and value "0" for all others. (There are other ways to transform a categorical variable into a set of numerical variables, but here we focus on the classic 0/1 dummy variable formulation.)

To illustrate the construction of dummy variables, suppose we have survey data on a sample of people and want to use race as an independent variable in a regression. Suppose also that in our data the survey respondents' racial self-identification has been recorded with five categories: African American, Asian, Native American, white, and other. Here are racial self-identifications for the first eight people in the sample:

Person ID	Race
1	African American
2	other
3	white
4	Asian
5	Native American
6	African American
7	white
8	white

To transform race into a set of dummy variables, we first need to choose a reference category. Suppose we picked "other," realizing that we could have picked any category as the reference without fundamentally changing the subsequent regression results. Then we need to create 5 – 1 = 4 dummy variables, corresponding to the four categories that were *not* chosen as the reference. These dummy variables will correspond to the categories African American, Asian, Native American, and white. For the African American dummy variable, African American respondents have value 1, and everyone else has value 0. For the Asian dummy variable, Asian respondents have value 1, and all non-Asian respondents have value 0. The Native American and white dummy variables are constructed in a similar way. The resulting values of the dummy variables for these eight subjects are shown below:

Person ID	Race	Dummy Variables for Race			
		"African American" Dummy	"Asian" Dummy	"Native American" Dummy	"White" Dummy
1	African American	1	0	0	0
2	other	0	0	0	0
3	white	0	0	0	1
4	Asian	0	1	0	0
5	Native American	0	0	1	0
6	African American	1	0	0	0
7	white	0	0	0	1
8	white	0	0	0	1

Notice that the 1's pick out the members of the categories named in the dummy variables. Also, person 2's race is "other," the reference category, so this person is not a member of any of the categories named in the four dummy variables. Therefore person 2 has value 0 for all the dummy variables. This shows how we have not lost any information on the respondents' race by leaving out a reference category, because we can still identify members of the reference category by looking for respondents with value 0 for all of the dummy variables in the set. In fact if we try to perform a regression without having left out a reference, statistical software will return an error message or perhaps notify us that it went ahead with the regression after omitting a dummy of its choosing. Advanced texts discuss the technical reasons why the method by which the software calculates the b's will not work if we fail to omit a reference category.

Then these four dummy variables will be included in the regression to represent the categorical variable race. Note that in calculating degrees of freedom (df) for the t-statistics in the regression output, each dummy variable counts as a separate X. In the current example, the dummy variables would count as 4 X's, not 1, with respect to df, even though conceptually they represent the single categorical independent variable race.

As mentioned above, we can choose any category as the reference without affecting regression results beyond some cosmetic, rather than fundamental, differences. The dummy variables we construct will convey the same information no matter which reference we use. What if we pick "African American" as the reference category? Then we will need to create dummy variables for the remaining categories: Asian, Native American, other, and white. We can illustrate this with the same eight people as above, again noticing that those in the reference category (now African American) have the value 0 for all of the dummies:

Person ID	Race	Dummy Variables for Race			
		"Asian" Dummy	"Native American" Dummy	"Other" Dummy	"White" Dummy
1	African American	0	0	0	0
2	other	0	0	1	0
3	white	0	0	0	1
4	Asian	1	0	0	0
5	Native American	0	1	0	0
6	African American	0	0	0	0
7	white	0	0	0	1
8	white	0	0	0	1

It is certainly possible for a regression model to include more than one categorical independent variable and therefore include more than one set of dummy variables (each with its own reference category, and with each set of dummy variables numbering one less than the number of categories in the corresponding categorical variable). For example, suppose that socioeconomic status is measured with three categories (upper, middle, and low), and is included in the regression along with a race variable that has five categories. In this case, the socioeconomic status variable will require two dummy variables and the race variable will require four, so the regression model as a whole will include a total of six dummy variables.

3.3 Interpretation of b for Dummy Variables

Interpretation of b for a dummy variable differs from interpretation of b for a numerical independent variable. For numerical independent variables, recall that we interpret b as the change in $\hat{Y}$ as X increases by one unit, holding other X's constant. But for dummy variables, we need to think about differences between categories, not "an additional unit of X," because the dummies give information on category membership, not the numerical value of some quantitative measurement. More specifically, a dummy variable's b indicates the difference in $\hat{Y}$ between the category named in the dummy and the reference category, controlling for, or holding constant, other X's. If this b is positive, then $\hat{Y}$ is higher (holding other X's constant) in the category represented by the dummy than in the reference category. If this b is negative, then $\hat{Y}$ is lower in the dummy's category than in the reference category.

When looking at regression results in the previous chapter, we considered statistical significance before looking at the b's, because usually we will only interpret b when there is statistical significance (or close to it). With dummy variables, we will still follow that approach of first assessing statistical significance before interpreting b's. But for the moment, as we are learning how dummy variables work, it is helpful to start by discussing the interpretation of b's before learning how to assess statistical significance.

We can continue with the above race example to show how b for a dummy variable indicates the difference in $\hat{Y}$ between the category named in the dummy variable and the reference category, holding other independent variables constant. Suppose we wanted to investigate how race and education predict television viewing habits. Then the regression will have race and education measures as independent variables and a measure of television viewing as the dependent variable. In the example above, the categorical race variable has five categories ("African American," "Asian," "Native American," "white," and "other"), and therefore leads to 5 – 1 = 4 dummies. If we decide to choose other as the reference category, we must create dummy variables for African American, Asian, Native American, and white. Education is measured as years of education, so that is just a numerical X as we have seen before. Television habits are measured as hours of viewing per week (Y). The regression equation is as follows:

$$\hat{Y} = a + b_1X_1 + b_2X_2 + b_3X_3 + b_4X_4 + b_5X_5$$

- X_1 = dummy variable for African American
- X_2 = dummy variable for Asian
- X_3 = dummy variable for Native American
- X_4 = dummy variable for white
- X_5 = education (years).

X_5 is a numerical independent variable, so the interpretation of b_5 will be as in the previous chapter. b_5 is the usual effect of an additional year of education on $\hat{Y}$, holding other independent variables (in this case, race) constant.

To help us understand the interpretation of the b's for the dummy variables (b_1, b_2, b_3, and b_4), we can rewrite the above regression equation separately for each racial group. That is, we will simplify the equation by plugging in the values of the dummy variables indicated for that group. For people in the reference category other, all dummy variables have value 0, so we rewrite the equation for the other group by setting X_1, X_2, X_3, and X_4 equal to 0:

$$\hat{Y} = a + b_1(0) + b_2(0) + b_3(0) + b_4(0) + b_5X_5$$

Therefore the rewritten equation for the other group is $\hat{Y} = a + b_5X_5$. We can do the same thing for African Americans, except that in this case $X_1 = 1$ and X_2, X_3, and $X_4 = 0$. This gives

$$\hat{Y} = a + b_1(1) + b_2(0) + b_3(0) + b_4(0) + b_5X_5,$$

and the rewritten equation for the African American group is $\hat{Y} = a + b_1 + b_5X_5$. Continuing the same way, we obtain the Asian equation by setting $X_2 = 1$ and X_1, X_3, and $X_4 = 0$. This means that our equation is

$$\hat{Y} = a + b_1(0) + b_2(1) + b_3(0) + b_4(0) + b_5X_5,$$

so the rewritten Asian equation is $\hat{Y} = a + b_2 + b_5X_5$. The Native American equation uses $X_3 = 1$ and X_1, X_2, and $X_4 = 0$, which leads to $\hat{Y} = a + b_3 + b_5X_5$. Finally, for whites we set $X_4 = 1$ and X_1, X_2, and $X_3 = 0$, and the resulting white equation is $\hat{Y} = a + b_4 + b_5X_5$.

First, compare the rewritten equations for the "African American" and "other" groups:

$$\begin{aligned} \text{African American}: \quad \hat{Y} &= a + b_1 + b_5X_5. \\ \text{Other}: \quad \hat{Y} &= a + b_5X_5. \end{aligned}$$

Because b_1, the b for the African American dummy variable, is the only difference between these two equations, b_1 represents the difference in $\hat{Y}$ between people—with the same level of education, because we are holding other X's constant—in the African American (the category named by the dummy) and other (the reference category) groups. If b_1 is positive, then predicted television viewing ($\hat{Y}$) is, holding education constant, higher in the African American than in the other group. If b_1 is negative, predicted television viewing ($\hat{Y}$) is, again holding education constant, lower in the African American than in the other group (or, equivalently, it is higher in the other than in the African American group).

The same kind of comparison shows that b_2, which is associated with the Asian dummy, represents the difference in $\hat{Y}$ (holding education constant) between Asians and people in the reference category other:

$$\begin{aligned} \text{Asian}: \quad \hat{Y} &= a + b_2 + b_5X_5. \\ \text{Other}: \quad \hat{Y} &= a + b_5X_5. \end{aligned}$$

Likewise, b_3, the b for the Native American dummy, is the difference in $\hat{Y}$ between Native Americans (the category this dummy represents) and those in the other category (the reference category), holding education constant:

$$\begin{aligned} \text{Native American}: \quad \hat{Y} &= a + b_3 + b_5X_5. \\ \text{Other}: \quad \hat{Y} &= a + b_5X_5. \end{aligned}$$

Finally, b_4 is the difference in $\hat{Y}$ between whites, the category named in the dummy associated with b_4, and people in the other category (the reference), again holding education constant:

$$\begin{aligned} \text{White}: \quad \hat{Y} &= a + b_4 + b_5X_5. \\ \text{Other}: \quad \hat{Y} &= a + b_5X_5. \end{aligned}$$

We can make this more concrete with some numbers. Suppose that b_1 (the b for the African American dummy) = –1. Then the difference in predicted television viewing between African Americans and people in the other category is –1 hours per week, controlling for education. The negative sign means that predicted TV viewing is 1 hour *less* for African Americans (the category named in the dummy) than for people in the other (the reference category) group, holding education constant; remember that a negative b indicates that $\hat{Y}$ is lower in the category named in the dummy than in the reference category. We could equally well say that predicted television viewing is 1 hour *more* for those in the other group than for African Americans, holding education constant.

Continuing with this example, if $b_{2,}$ associated with the Asian dummy, equals 3, then predicted television viewing for Asians is 3 hours per week more than that for members of the other category, holding education constant. Suppose that b_3, the b for the Native American dummy variable, equals 1. That means that predicted viewing is 1 hour per week greater for Native Americans than for people in the other category, controlling for education. Finally if b_4, the b for the white dummy, equals 6, this indicates that predicted TV viewing is 6 hours per week higher for whites than for those in the other category, holding education constant.

As with b's for regular numerical variables, we also need to consider real-world significance, deciding whether the differences in $\hat{Y}$ between categories are large enough to be meaningful. (Keep in mind that when we really look at our regression results, we will assess statistical significance before looking at the b's, but again we are deferring our discussion of statistical significance with dummies for the moment.)

Especially when there are many categories, it may be helpful to display the b's from the set of dummies (representing a single categorical independent variable) on a number line. Then it is easy to see the order and spacing of the categories in terms of predicted Y. $\hat{Y}$ increases as we move from left to right, and we always put the reference category at the 0 point on the number line. We mark both the category and the value of b on the line and can then see the spacing of the adjacent categories:

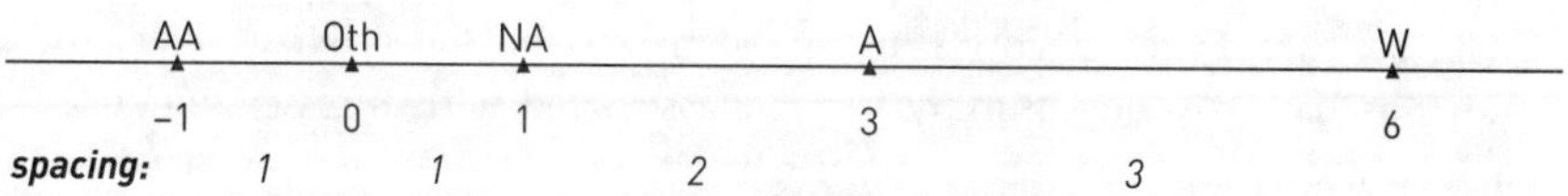

From the number line, it is easy to see that, controlling for education, our analysis estimates that whites watch TV most, followed by Asians, Native Americans, and those in the other group, with African Americans watching least. It is also easy to make not only the comparisons of each category to the reference (given by the b's), but also any comparisons of two non-reference categories. For example, we see that the difference in $\hat{Y}$ (predicted TV viewing) between whites and Native Americans is 6 – 1 = 5, so predicted viewing is 5 hours per week greater for whites than for Native Americans, again holding education constant. Comparing Asians and African Americans, the difference is 3 – (–1) = 4, so predicted viewing is 4 hours per week more for Asians than for African Americans.

The number line also lets us quickly determine, without running another regression, what the b's would be if we changed the reference category. Suppose that we used Asian as the reference category instead of other. Then, as the reference, Asian will be placed at 0 on the number line. To find the values for other categories in the number line, which are the new b's obtained when using Asian as the reference, we use the fact that the order and spacing of the categories in terms of $\hat{Y}$ is unaffected by the choice of reference category. So if "Asian" is now 0 in the number line, then "white" should be 3, "Native American" should be −2, "other" should be −3, and "African American" should be −4. The new number line (with the values for the points indicating the b's) is therefore:

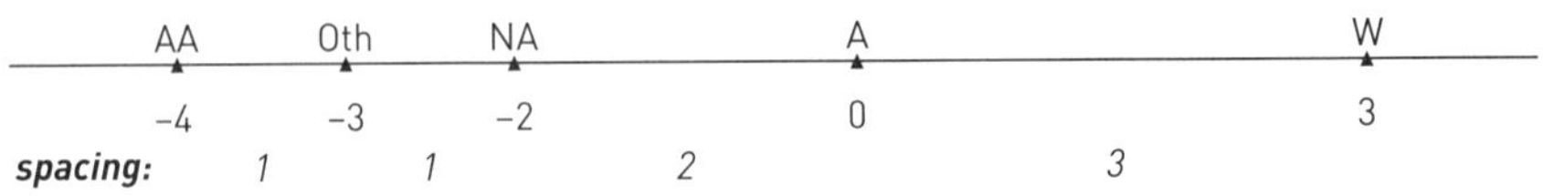

This number line gives exactly the same information about order and spacing of the categories in terms of predicted television viewing (Y) as the previous number line, but now the numerical labels, corresponding to the new b's we would obtain, indicate differences from the new reference category Asian.

3.4 Statistical Significance and Dummy Variables

The previous section showed how to interpret b's for dummy variables and how the interpretation differs from that involving numerical independent variables. In discussing that interpretation, we did not address statistical significance, even though in practice we will think about statistical significance for dummy variables before attempting to interpret the b's, just as we do with numerical variables. We now discuss statistical significance in the context of dummy variables.

3.4.1 Overall Assessment of Statistical Significance of a Categorical Independent Variable

First note that if the categorical independent variable has only two categories, and thus has only one dummy variable associated with it, assessing statistical significance is no different from what it was for a numerical variable. In that case, we can simply follow the usual procedure by looking at the p-value obtained from the t-statistic for the dummy variable and comparing it to the 0.05 cutoff. If the p-value is small (that is, under the cutoff), we decide that the difference in the dependent variable between the two categories is statistically significant. For example, some data sets include just two categories for employment status, "Employed" or "Unemployed," so only one dummy variable is needed. Suppose we chose "Unemployed" as the reference category for employment status, and therefore have "Employed" as the dummy variable. If the p-value for the "Employed" dummy variable is less than or equal to 0.05, the difference in $\hat{Y}$ between employed and unemployed is statistically significant. We can then interpret the b for the "Employed" dummy variable as the difference in $\hat{Y}$ between employed and unemployed, controlling for other independent variables in the regression. As with numerical variables, we will resist making a further interpretation of b when the difference is not statistically significant, although as before we will be sensitive to "borderline" p-values.

Assessing statistical significance is more complicated when the categorical X has more than two categories, because in that case we need to create multiple dummy variables and include all of them in the regression analysis. Each dummy will have a p-value from a t-statistic as before, but these refer to the significance of the difference in the dependent variable between that dummy's category and the reference. Because we want to assess the statistical significance of the entire categorical X, it is not clear how we should try to combine these p-values into a single conclusion. For instance, what would we conclude if some of the dummies in a set representing a single categorical independent variable show p-values less than our 0.05 cutoff, but others from that set have p-values greater than the cutoff? A further difficulty is that these p-values will change when we change the reference category, as b's, standard errors, and t-statistics change with different choices of the reference. (Thinking of the number lines above, our regression output is more apt to show statistically significant differences from the reference category when the reference is at one end of the number line than when it is in the middle.) For these reasons, we cannot use the usual p-values, derived from t-statistics, on the regression output to assess statistical significance of a categorical independent variable that contains more than two categories and yields more than one dummy variable.

Rather than relying on this collection of p-values from t-statistics, we will use a method that considers the entire set of dummy variables associated with the categorical independent variable. This method asks whether the regression performs significantly better (in the sense of how well the predicted values of Y match the actual values in our sample data) when the set of dummies is included among the independent variables than it does when the set is not included. We can think of this intuitively as a comparison of R^2 for the regression with and without the set of dummy variables.

Of course we said in the last chapter that R^2 will always be at least a little higher when we add more independent variables to the regression. But in assessing the statistical significance of the set of dummy variables, we want to know whether the difference in R^2 was sufficient to justify focusing on the more complicated—that is, including more independent variables—model. The method weighs the difference in R^2 against the additional df used in the model that includes the dummy variables. If the difference in R^2 seems large relative to the additional df used, we conclude that the categorical independent variable has a statistically significant impact on Y. If the difference in R^2 seems small relative to the additional df used, then we conclude that the categorical independent variable does not have a statistically significant impact on Y. In this way we will have an overall test for the set of dummy variables together, rather than somehow attempting to combine the dummies' individual p-values.

This comparison of R^2 for the model with and without the set of dummy variables is formalized via an *F-test*. Like a t-test, the F-test involves calculating a particular test statistic (the *F-statistic*) and then using a table (or an online probability calculator, or statistical software) to obtain the p-value corresponding to the value of the test statistic. The F-test can actually be applied in a variety of situations in regression beyond that of assessing statistical significance for a set of dummy variables; the Appendix to this chapter discusses some of the other uses of the F-test (see Sections 3.8.1–3.8.3).

For the situation that we are considering here, the null hypothesis (H_0) for the F-test is that in the true regression model—what we would have if our data included the whole population (or superpopulation) instead of a sample—the β's for the set of dummy variables are all zero. Because those β's represent the differences between categories (holding other independent variables constant), if all are zero then there are on average no differences in Y between categories, and the categorical variable is of no use in predicting Y. The alternative hypothesis (H_1) is that at least one β is not zero, meaning that there is at least some difference in predicted Y among these categories; this would not necessarily mean that *all* of the categories are different—just that in some way they are not all the same with respect to Y. If this alternative hypothesis (H_1) is correct, then the categorical variable is useful in making predictions in the population. With this setup, rejecting the null hypothesis (that the β's for the set of dummy variables are all zero) means that the alternative hypothesis (that at least one β is not zero) is more reasonable, and we would interpret that result as showing a statistically significant impact of the categorical independent variable on Y.

Unlike the t-statistics discussed in Chapter 2 for testing the hypothesis that $\beta = 0$, the F-statistic that we need for the set of dummies will not in general show up on our regression output. In fact there is an F-statistic, and its associated p-value, on the default output produced by all of the standard statistical software packages, but it is usually not what we want for testing the set of dummy variables. (Section 3.8.2 in the Appendix to this chapter explains what the F-statistic on the output represents.) Although some software can be given commands to calculate the F-statistic that we do want, it is actually not too difficult to calculate ourselves, and we show that here.

To calculate F, we need the regression output for two different models. We will denote the regression that includes the set of dummy variables we are testing, plus whatever other X's are in the regression, as "Model 1" and use the output from that model in calculating F. We also need the output from what we will call "Model 0," which includes the other X variables we are using but does **not** include the set of dummy variables being tested. Sometimes Model 1 is referred to as the "full model" in this situation, to highlight that the set of dummy variables is being excluded from Model 0, or the "larger model," because it includes more independent variables.

From the software output for these two regression analyses, we can find the "sum of squared errors" for each. This number is obtained by squaring the usual errors ($Y - \hat{Y}$), the differences between the actual and predicted Y for each case, and summing over all cases in the sample. In some software this will be labeled the "sum of squared residuals," as residual is another name for the error ($Y - \hat{Y}$). We label the sum of squared errors from Model 1 (the full model, including all the X's) as SSE_1, and the sum of squared errors from Model 0 (without the set of dummy variables being tested) as SSE_0. Also, let df_1 represent the df for t in Model 1, so that $df_1 = N - (\#\text{ of X's in Model 1}) - 1$, and let "s" represent the number of dummy variables in the set we are testing. Then we calculate the F-statistic (referred to as F) as

$$F = \frac{(SSE_0 - SSE_1)/s}{SSE_1/df_1}$$

Because df_1 will appear in the denominator of the F-statistic, it may be called "denominator degrees of freedom." Likewise, "s" may be called "numerator degrees of freedom." We mentioned above that the F-test is intuitively a comparison of R^2 between Model 0 and Model 1, and this formula for F is consistent with this intuition. The R^2 for a single model is a function of SSE, with a higher R^2 value obtained when SSE is smaller (indicating a closer fit between the actual and predicted values of Y in the sample). The F formula thus embodies the idea of comparing the two models' R^2 values, with this comparison being made relative to R^2 for Model 1, which has the additional set of dummy variables. Writing R_0^2 (R^2 for Model 0) and R_1^2 (R^2 for Model 1) for the two models' R^2 values, it can be shown that the F formula above is equivalent to

$$F = \frac{(R_1^2 - R_0^2)/s}{(1 - R_1^2)/df_1}$$

The F-statistic cannot be negative. First, going back to the earlier expression for F above, the denominator (SSE_1 / df_1) is based on a sum of squared numbers: any number squared is positive, and a sum of positive numbers is also positive. Because df_1 is also positive, dividing a sum of positive numbers by df_1 gives a positive denominator in the F-statistic. Second, and a bit less plainly, the difference in the numerator [(SSE_0 – SSE_1) / s] cannot be negative. That is because Model 1 includes all the variables in Model 0, plus additional ones (the set of dummies being tested). In that situation, Model 1 will not make worse predictions overall than Model 0, because its additional independent variables allow it more flexibility in trying to make its predictions match the actual Y values in the sample. (Model 1's $\hat{Y}$ could be farther than Model 0's from the actual Y for some particular cases in our sample, but the overall pattern will be that of smaller errors under Model 1.) This means that Model 0 will have a larger SSE (and a smaller R^2) than Model 1, so that (SSE_0 – SSE_1) will not be negative. s is also a positive number, and the result is that the numerator of the F-statistic will be positive. With a positive numerator and denominator, the F-statistic will be positive.

Our goal is to find the p-value for the null hypothesis of no average differences in the dependent variable between any categories of the categorical independent variable being tested, so we need to look up this F-statistic in an F-table (page 237). Finding a p-value from the F-table also requires the two different df mentioned above. The first, referring to the numerator of the F formula above, is s (or numerator degrees of freedom). The second refers to the denominator of the F formula, and is df_1 (or denominator degrees of freedom).

If the p-value that we determine from the table is $\leq$ 0.05, we conclude that the categorical independent variable has a statistically significant impact on the dependent variable. When that happens, our improved accuracy in predicting Y in the sample, as indicated by ($SSE_0 - SSE_1$), when including the set of dummies is substantial enough to be worth the complication of the additional variables being used in the regression. If the p-value is > 0.05, then we conclude that the categorical independent variable does not have a statistically significant impact on the dependent variable, and the improvement in the accuracy of the predictions is not great enough to justify including the set of dummy variables. For example, if $F = 2.84$, s (numerator degrees of freedom) = 3, and df_1 (denominator degrees of freedom) = 16, then the table shows that the p-value falls between 0.05 and 0.10. Because this exceeds our 0.05 cutoff, we conclude that the categorical independent variable does not have a statistically significant impact on Y. (A more precise p-value for the F-statistic can be obtained from an online calculator or in statistical software, and using one of those tells us that $p = 0.071$; printed F-tables will not allow that much precision.)

3.4.2 Assessment of Statistical Significance in the Difference in $\hat{Y}$ Between Two Categories

The p-value from the F-statistic leads to our overall assessment of statistical significance for the categorical independent variable represented in the set of dummies. There may also be research purposes for which we are interested in comparisons of $\hat{Y}$ between specific categories of the categorical independent variable. For example, after finding overall statistical significance of a race variable via the F-test, the researcher may be interested in assessing the statistical significance of the difference in $\hat{Y}$ between two particular racial categories, such as between whites and Asians, or between African Americans and Native Americans, and so on. Perhaps we are considering whether it would be reasonable to combine certain categories of this independent variable, or some theory is especially focused on two particular categories. Note that even if we are interested in exploring such details, as discussed in the paragraphs to follow, our summary statement of the categorical variable's statistical significance will still be based on the F-test.

When examining the difference in $\hat{Y}$ between two specific categories of the independent variable, we use a t-test to assess statistical significance of the difference. Our software output will give a p-value for the usual t-test for each dummy variable in the model, and from such a p-value we can immediately assess the statistical significance of the difference in the dependent variable between a category named in a dummy variable and the reference category.

Using the example of race above (with five categories "African American," "Asian," "Native American," "white," and "other"), suppose that we chose "Asian" as the reference category. Then the p-values for the dummy variables in the output will refer to hypotheses of no difference in $\hat{Y}$ between each category named in a dummy variable and the reference category. In this particular example, we can look at the p-value for the other dummy to decide if there is a statistically significant difference in $\hat{Y}$ between "other" and "Asian." Similarly, we can look at the p-value for the African American dummy to assess the statistical significance of the difference between "African American" and "Asian." Likewise, the p-value for the white dummy indicates if there is a statistically significant difference in $\hat{Y}$ between "white" and "Asian." Finally, we use the p-value for the Native American dummy to evaluate statistical significance of the difference between "Native American" and "Asian."

Note, however, that when the categorical independent variable has more than two categories, like in this example of race, the set of displayed p-values in the output will not cover every possible comparison between categories. Only those comparisons involving the chosen reference category will have a displayed p-value. That means that the displayed p-values will not allow us to assess the statistical significance of the difference in $\hat{Y}$ between two non-reference categories; with "Asian" as reference, this includes comparisons such as "other" and "Native American," "African American" and "white," and so on. The easiest way to obtain p-values for those other comparisons is to simply change the reference category as needed and rerun the regression using the new choice of reference category. As discussed above, the choice of reference category is arbitrary, and does not fundamentally affect the results of the regression analysis. But one cosmetic difference after changing the reference category is that the p-values for the dummy variables shown in the output will highlight different comparisons between categories than before. In the example above, switching the reference category to "other" would provide p-values for the hypothesis of no difference in Y between the following pairs of categories: (a) "African American" and "other," (b) "Native American" and "other," (c) "white" and "other," as well as (d) the "Asian" and "other" comparison for which we would have already had a p-value from the analysis using "Asian" as reference. We can switch the reference category as needed to obtain a p-value and assess the statistical significance of the difference in $\hat{Y}$ between any two categories.

3.5 Finding Predicted Values of Y in Multiple Regression With Dummy Variables

The process of obtaining predicted values of Y from a multiple regression with dummy variables is really no different from what we saw in the previous chapter for regression with numerical independent variables. For example, suppose we had survey data on adults who had completed their formal education, and that we wanted to examine how age, education (categorized by highest degree as "High School or Less," "College," and "Graduate Degree"), and employment status ("Employed" and "Unemployed") affect alcohol consumption (measured by average number of drinks per week). Suppose also that we selected "High School or Less" as the reference category for education and "Unemployed" as the reference category for employment status.

Then our regression model would include age (X_1) and dummy variables for "College" (X_2) and "Graduate Degree" (X_3), representing education, and "Employed" (X_4), representing employment status. Suppose that we obtained results showing that the b for age was 0.043, the b for the "College" dummy variable was 1.4, the b for the "Graduate Degree" dummy variable was –1.2, and the b for the "Employed" dummy variable was –2.4. Also, we obtained a Y-intercept of 2.71. Then the estimated regression equation from this analysis would be

$$\hat{Y} = 2.71 + 0.043X_1 + 1.4X_2 + (-1.2)X_3 + (-2.4)X_4, \text{ where}$$

- X_1 = age
- X_2 = "College" dummy variable
- X_3 = "Graduate Degree" dummy variable
- X_4 = "Employed" dummy variable.

What is the predicted alcohol consumption from this regression equation for an employed person who is 26 years old and has graduated from college but does not have any more advanced degree? Such a person has the value 26 for age (X_1), 1 for the "College" dummy variable (X_2), 0 for the "Graduate Degree" dummy variable (X_3), and 1 for the "Employed" dummy (X_4). Plugging those values into regression equation, we have

$$\hat{Y} = 2.71 + 0.043(26) + 1.4(1) + (-1.2)(0) + (-2.4)(1) = 2.83$$

Therefore, 2.83 drinks per week is the predicted alcohol consumption from this regression for an employed person who is 26 years old and whose highest degree is a bachelor's.

What is the predicted alcohol consumption for an unemployed person who is 24 years old and has only a high school diploma? For this person, we plug in 24 for age, 0 for the "College" dummy variable, 0 for the "Graduate Degree" dummy variable, and 0 for the "Employed" dummy:

$$\hat{Y} = 2.71 + 0.043(24) + 1.4(0) + (-1.2)(0) + (-2.4)(0) = 3.74$$

The regression model predicts 3.74 drinks per week for a 24-year-old unemployed person who has only a high school diploma. Remember that, as before, we need to consider the value of R^2 and the range of X values in the sample being used to estimate the multiple regression model when deciding how much faith we should put in this prediction.

3.6 Example: Robbery Rates in American States

This example will give us practice in creating dummy variables from categorical independent variables, running regression analysis with dummy variables, and interpreting results. A criminologist hypothesized that greater urbanization and unemployment increase crime in society. She also believed that, even after controlling for other factors, the Southern region of the United States would have more crime than other regions. Using 29 American states as her sample, the researcher obtained measurements of crime and urbanization, unemployment, and region from the U.S. Census Bureau and the FBI's Uniform Crime Reports. (For convenience, this example used a sample of only 29 states, but for an actual research project involving state data we would typically use all 50.)

Crime was measured by robbery rate, defined for each state as the number of robberies reported to police per 100,000 population. Urbanization was measured as the population density—specifically, population per square mile—in each state, and unemployment was measured as the percentage of the labor force that was unemployed. Region distinguished among three areas of the United States: "East," "South," and "Midwest."

The research hypothesis, units of analysis, measurements of independent and dependent variables, and raw data are as follows:

Research hypothesis: Greater urbanization and unemployment increase a society's crime rate. The South has a higher level of crime than other regions of the United States.

Units of analysis: 29 U.S. states (aggregate-level data)

Measurements of independent and dependent variables:

Y: Robberies reported to police per 100,000 population (crime).

X_1: Population per square mile (urbanization).

X_2: Percentage of labor force that was unemployed (unemployment).

X_3: Region: "East," "South," "Midwest" (region).

Research hypothesis with independent and dependent variable measures:

1. Population per square mile (X_1) has a positive effect on the number of robberies per 100,000 population (Y).
2. The percentage of the labor force that is unemployed (X_2) has a positive effect on the number of robberies per 100,000 population (Y).
3. States located in the South have higher robbery rates (Y) than states located in other regions.

Data:

State ID	State Name	Robbery Rate per 100,000 (Y)	Population per Square Mile (X_1)	Percentage Unemployed (X_2)	Region (X_3)
1	Alabama	86.5	96.26	5.8	South
2	Delaware	112.5	493.56	5.3	East
3	Florida	88.6	391.48	5.5	South
4	Georgia	96.3	181.38	5.8	South
5	Illinois	137.2	230.59	6.1	Midwest
6	Iowa	39.8	56.32	3.6	Midwest

(Continued)

State ID	State Name	Robbery Rate per 100,000 (Y)	Population per Square Mile (X_1)	Percentage Unemployed (X_2)	Region (X_3)
7	Kansas	61.3	35.63	4.2	Midwest
8	Kentucky	66.4	112.79	5.5	South
9	Louisiana	114.4	108.44	6.5	South
10	Maine	18.6	43.32	4.2	East
11	Massachusetts	71.0	879.35	4.6	East
12	Michigan	65.1	176.24	5.9	Midwest
13	Minnesota	64.9	70.05	3.6	Midwest
14	Mississippi	69.4	63.60	7.0	South
15	Missouri	103.9	88.97	4.6	Midwest
16	Nebraska	50.4	24.99	3.3	Midwest
17	New Hampshire	31.2	150.00	3.8	East
18	New Jersey	87.7	1,224.59	5.3	East
19	North Carolina	91.0	211.30	5.3	South
20	North Dakota	24.2	10.95	2.9	Midwest
21	Ohio	99.5	285.34	5.2	Midwest
22	Pennsylvania	92.1	286.23	5.3	East
23	South Carolina	77.0	167.09	5.8	South
24	South Dakota	27.4	11.47	3.5	Midwest
25	Tennessee	117.1	162.87	4.9	South
26	Texas	114.0	108.35	5.1	South
27	Vermont	14.8	67.66	3.8	East
28	West Virginia	28.9	75.54	6.7	East
29	Wisconsin	75.0	107.02	3.5	Midwest

Because region is a categorical variable, the researcher needed to create dummy variables, first selecting a reference category and then creating dummy variables for the remaining categories. She selected "East" as the reference category and created dummy variables for "South" and "Midwest." The resulting dummy variables look like this:

State ID	State Name	Region (X_3)	Dummy Variables for Region	
			"South" Dummy (X_{3a})	"Midwest" Dummy (X_{3b})
1	Alabama	South	1	0
2	Delaware	East	0	0
3	Florida	South	1	0
4	Georgia	South	1	0
5	Illinois	Midwest	0	1
6	Iowa	Midwest	0	1
7	Kansas	Midwest	0	1
8	Kentucky	South	1	0
9	Louisiana	South	1	0
10	Maine	East	0	0
11	Massachusetts	East	0	0
12	Michigan	Midwest	0	1
13	Minnesota	Midwest	0	1
14	Mississippi	South	1	0
15	Missouri	Midwest	0	1
16	Nebraska	Midwest	0	1
17	New Hampshire	East	0	0
18	New Jersey	East	0	0
19	North Carolina	South	1	0
20	North Dakota	Midwest	0	1
21	Ohio	Midwest	0	1
22	Pennsylvania	East	0	0
23	South Carolina	South	1	0
24	South Dakota	Midwest	0	1
25	Tennessee	South	1	0
26	Texas	South	1	0
27	Vermont	East	0	0
28	West Virginia	East	0	0
29	Wisconsin	Midwest	0	1

The South and Midwest dummy variables together represent the categorical variable region; in listing the variables, let us emphasize that by labeling those dummies as X_{3a} and X_{3b}, rather than X_3 and X_4. With that notation, the X's in the regression analysis will include population per square mile (X_1), the percentage unemployed (X_2), the South dummy variable (X_{3a}), and the Midwest dummy variable (X_{3b}). The regression equation is

$$\hat{Y} = a + b_1X_1 + b_2X_2 + b_3X_{3a} + b_4X_{3b}.$$

As with a regression involving only numerical X's, it is valuable to first look at correlations among the X's. Although the idea of a correlation seems somewhat strange when a dummy variable is involved, from the point of view of the regression analysis it is a correlation just the same. The correlations between dummies from the same set are not informative, but correlations between those dummies and other independent variables can be. For these data, the results of the correlation analysis are as follows:

	Population per Square Mile (X_1)	**Percentage Unemployed (X_2)**	**"South" Dummy (X_{3a})**	**"Midwest" Dummy (X_{3b})**
Population per Square Mile (X_1)	1.000	0.186	−0.122	−0.314
Percentage Unemployed (X_2)	0.186	1.000	0.538	−0.505
"South" Dummy (X_{3a})	−0.122	0.538	1.000	−0.567
"Midwest" Dummy (X_{3b})	−0.314	−0.505	−0.567	1.000

The highest (in absolute value) correlation is −0.567, between "South" dummy (X_{2a}) and "Midwest" dummy (X_{3b}), within the set of region dummies. None of the other correlations are too remarkable. Note that negative correlations among the set of dummy variables representing a single categorical variable make sense, because if a state has a "high" value of 1 for one dummy, it must have a "low" value of 0 for any of the other dummies associated with that categorical variable. It is often the case that pairs of dummies representing categories of the same independent variable show rather high negative correlations, but again we will not think of those as indicating possible collinearity problems.

Regression With "East" as Reference Category

Because "East" is the reference category, the multiple regression model included population per square mile (X_1), the percentage unemployed (X_2), the "South" dummy (X_{3a}), and the "Midwest" dummy (X_{3b}). The regression analysis output looks like this:

$R^2 = 0.522$
df for t = 24

	b	SE of b	t	p-value
Intercept	−27.320	27.377	−0.998	0.328
Population per square mile (X_1)	0.059	0.021	2.834	0.009
Percentage unemployed (X_2)	12.475	5.383	2.318	0.029
"South" dummy (X_{3a})	38.630	13.917	2.776	0.011
"Midwest" dummy (X_{3b})	36.911	13.282	2.779	0.010

The regression output shows first that R^2 was only moderate; this suggests that there are other important predictors of state robbery rates that have not been included in this regression and that in this sample the predicted robbery rates from this regression will not necessarily be very close to the actual rates. Only about 52% of the variance in robbery rates was explained by this set of X's. The output also shows that the effect of population per square mile (X_1) on robbery rate (Y) was statistically significant, controlling for the percentage unemployed (X_2) and region (X_3). The effect of the percentage unemployed (X_2) on robbery rate (Y) was also statistically significant, controlling for population per square mile (X_1) and region (X_3).

Both p-values for dummy variables representing region (X_3) were < 0.05 (p = 0.011 for the "South" dummy, and 0.010 for the "Midwest" dummy), indicating statistically significant crime differences in at least two of the three pairs of regions. Remember, however, that in general when (as here) a categorical independent variable requires more than one dummy, we should conduct an F-test to make an overall assessment of statistical significance. That test requires the sum of squared errors from this regression (SSE_1, from Model 1) as well as from another regression that excludes the dummy variables being tested (SSE_0, from Model 0). The full output from statistical software for the above model that includes population density, unemployment, and the region dummies reports that SSE_1 is 14,822.625, though this is not shown in the summary above. Note that SSE_1, like R^2, will be the same no matter which reference category is used for region in Model 1.

To obtain SSE_0, the researcher needed to run another regression (Model 0) that excludes the set of dummy variables being tested. In the present example, Model 0 would include only population per square mile (X_1) and percentage unemployed (X_2). From the output for this regression leaving out the region dummies, $SSE_0 = 21{,}124.834$. To calculate F, the researcher also needed s, the number of dummy variables in the set being tested, and df_1, the df for t in Model 1 that includes the set of dummies. In this example s = 2, because region had three categories and, thus, two dummies associated with it. $df_1 = N - (\#\ X\text{'s in the full model}) - 1 = 29 - 4 - 1 = 24$. (As noted above, each dummy counts as an X when calculating df.) Then F is calculated as

$$F = \frac{(SSE_0 - SSE_1)/s}{SSE_1/df_1}$$

$$F = \frac{(21,124.834 - 14,822.625)/2}{14,822.625/24}$$
$$= \frac{6,302.209/2}{14,822.625/24} = \frac{3,151.105}{617.609} = 5.102.$$

The F-table with (2, 24) df is used to find the p-value, because s (numerator df) = 2 and df_1 (denominator df) = 24. The table (page 237) shows that the p-value corresponding to F = 5.102 lies between 0.010 and 0.025 (Statistical software or an online calculator gives the precise p-value of 0.014). This p-value is less than the 0.05 cutoff, so the regression with the two region dummies does fit the sample data significantly better than the regression without them, and the researcher interpreted this as showing that there is a statistically significant impact of region.

Having found statistical significance for population per square mile (X_1), percentage unemployed (X_2), and region (X_3), interpretation of the b's for these variables is appropriate. The predicted robbery rate per 100,000 (Y) increases by 0.059 as population per square mile (X_1) increases by 1 person, controlling for other X's. In light of the range (10.95–1,224.59) and mean (204.186) of population per square mile (X_1), it may help with interpretation to instead consider an increase of 10 people per square mile: the predicted robbery rate per 100,000 (Y) increases 0.59 as population per square mile (X_1) increases by 10 people. Because robbery rate (Y) has a mean of 73.317 (and range of 14.8–137.2), this increase of 0.59 in the predicted robbery rate per 100,000 (Y) still seems rather small in magnitude. The predicted robbery rate per 100,000 (Y) increases 12.475 as the percentage of unemployed persons (X_2) increases by 1%, controlling for other X's. Considering the descriptive statistics for the percentage unemployed (X_2) and robbery rate (Y), this estimated effect seems meaningful in real terms.

For the set of dummies, it is helpful to place their b's (and the spacing between adjacent categories) on the number line:

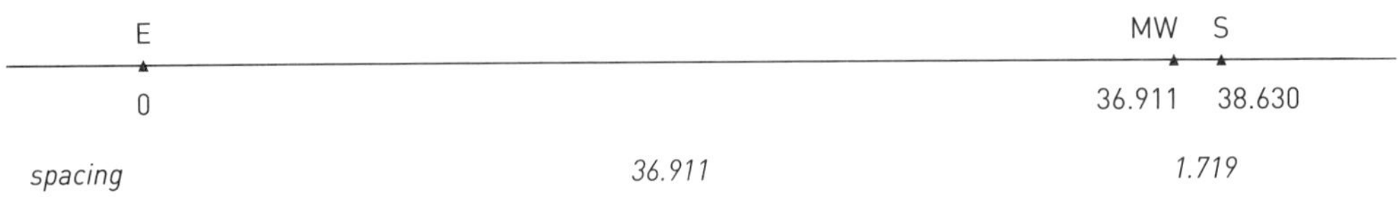

From b for the "South dummy" (X_{3a}), the difference in predicted robbery rate per 100,000 (Y) between "South" and the reference "East" is 38.630, controlling for population per square mile (X_1) and the percentage unemployed (X_2). In other words, the predicted robbery rate for a Southern state is 38.630 greater (per 100,000 population) than in an Eastern state, holding population per square mile (X_1) and the percentage unemployed (X_2) constant. In the same way, the difference in predicted robbery rate per 100,000 (Y) between "Midwest" and "East" is 36.911, controlling for other X's.

The final difference in predicted robbery rate is between "South" and "Midwest," as that is the third pair of regions. Neither of these regions was selected as the reference category, so this difference is not shown by any b in the output. However, it is easy to see from the number line: the predicted robbery rate (holding other X's constant) is 1.719 greater in Southern than in Midwestern states, calculated as $b_3 - b_4$ = 38.630 – 36.911. When interpreting these b's in light of the mean (73.317) and range (14.8–137.2) of robbery rates (Y) for the states, the researcher decided that the differences in predicted robbery rates between East and the other two regions seem large enough to have real-world significance, but the difference between the South and the Midwest is much smaller in magnitude, and may not be important in real terms. From the number line, she also concluded that after controlling for population per square mile (X_1) and unemployment (X_2), the Southern states have the highest predicted robbery rate (Y), followed closely by the Midwestern states. Eastern states had the lowest predicted robbery rate.

This interpretation covered all three possible differences in predicted robbery rates between regions, as the F-test indicated overall statistical significance of the region variable. If we want to focus on the significance of a particular difference between categories, we can use the t-test for the dummy showing that comparison. In this example with "East" as reference category, both of the p-values for dummy variables representing region (X_3) were < 0.05 ($p = 0.011$ for the "South dummy", and 0.010 for the "Midwest" dummy). These indicate that the South vs. East and the Midwest vs. East differences in predicted robbery rate were statistically significant.

What if the researcher wanted to assess statistical significance for the comparison between South and Midwest? Neither was used as the reference in this regression, so the appropriate t-test is not shown on the output above. But this is not really a problem, because it easy to obtain the p-value for a comparison of $\hat{Y}$ for two non-reference categories: simply rerun the regression analysis with one of those categories selected as the reference.

Regression With "Midwest" as Reference Category

Suppose that the researcher decided to use "Midwest" as the reference category. Then "South" (X_{3a}) and "East" (X_{3c}) dummy variables need to be included in the regression analysis to represent the categorical variable region (X_3). The dummy variables for "South" (X_{3a}) and "East" (X_{3c}) are shown below:

State ID	State Name	Region (X_3)	Dummy Variables for Region	
			"South" Dummy (X_{3a})	"East" Dummy (X_{3c})
1	Alabama	South	1	0
2	Delaware	East	0	1
3	Florida	South	1	0
4	Georgia	South	1	0
5	Illinois	Midwest	0	0
6	Iowa	Midwest	0	0
7	Kansas	Midwest	0	0
8	Kentucky	South	1	0
9	Louisiana	South	1	0
10	Maine	East	0	1

(Continued)

State ID	State Name	Region (X_3)	Dummy Variables for Region	
			"South" Dummy (X_{3a})	"East" Dummy (X_{3c})
11	Massachusetts	East	0	1
12	Michigan	Midwest	0	0
13	Minnesota	Midwest	0	0
14	Mississippi	South	1	0
15	Missouri	Midwest	0	0
16	Nebraska	Midwest	0	0
17	New Hampshire	East	0	1
18	New Jersey	East	0	1
19	North Carolina	South	1	0
20	North Dakota	Midwest	0	0
21	Ohio	Midwest	0	0
22	Pennsylvania	East	0	1
23	South Carolina	South	1	0
24	South Dakota	Midwest	0	0
25	Tennessee	South	1	0
26	Texas	South	1	0
27	Vermont	East	0	1
28	West Virginia	East	0	1
29	Wisconsin	Midwest	0	0

Here are the regression results with "Midwest" as the (excluded) reference category:

$R^2 = 0.522$
$df for t = 24$

	b	SE of b	t	p-value
Intercept (constant)	9.591	23.612	0.406	0.688
Population per square mile (X_1)	0.059	0.021	2.834	0.009
Percentage unemployed (X_2)	12.475	5.383	2.318	0.029
"South" dummy (X_{3a})	1.719	13.448	0.128	0.899
"East" dummy (X_{3c})	−36.911	13.282	−2.779	0.010

As discussed earlier, changing the reference category for region did not change R^2 or the results involving other variables. That is, the b's, standard errors, t-statistics, and p-values for population per square mile (X_1) and percentage unemployed (X_2) are exactly the same as in the previous output. Also, the order and spacing of the regions in the number line did not change, just the numerical labels and the marking of the new reference category as 0.

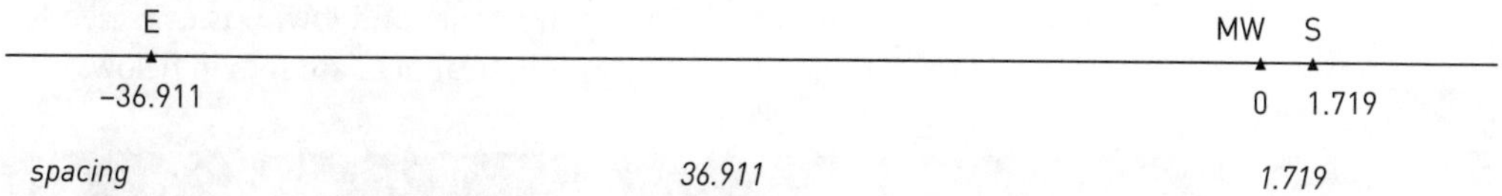

The F-test for the overall assessment of the region variable's statistical significance also remained the same. The new information in this output concerns the "South" vs. "Midwest" comparison, as there is now a p-value from a t-test to determine statistical significance of that difference. The South dummy's p-value is 0.899, or > 0.05, so the "South" vs. "Midwest" difference was not statistically significant. Although the overall assessment from the F-test indicated statistical significance of the region variable (X_3), that finding does not necessarily imply that all possible pairs of regions show a statistically significant difference in robbery rates. The overall statistical significance of the region variable is being driven by the large estimated difference between the East and the other two regions, not the much smaller estimated difference between the South and the Midwest. It is logical that overall statistical significance of the categorical independent variable would imply that at least one of the particular differences in $\hat{Y}$ between categories be statistically significant, but the reverse is not true. The presence of a statistically significant difference between two categories does not necessarily mean that the F-test will show statistical significance for the categorical variable overall, especially as the number of categories grows larger.

With these regression results, the researcher was able to evaluate her research hypothesis. Based on the earlier examination of p-values and slopes (b's), population per square mile (X_1) showed a statistically significant effect on robbery rates (Y) in the positive direction (robbery rate increases as population per square mile increases) that was suggested by the research hypothesis. However, the estimated effect of population per square mile (X_1) did not seem large enough to have much real-life significance. The effect of percentage unemployed (X_4) was statistically (and real-life) significant in the positive direction (robbery rate increases as percentage unemployed increases), as suggested by the research hypothesis. The overall impact of region (X_3) was statistically significant; examination of b's on the number line showed that the Southern states had the highest predicted robbery rates (controlling for the other X's), although not significantly higher than Midwestern states (with the estimated difference also quite small). R^2 is only moderate, however, indicating that it would be worthwhile in further analyses to explore additional X's that might also be related to robbery rates (Y).

3.7 Exercises

3.7.1 Exercise 1: Sentence Length for Convicted Robbery Offenders

Information on severity of punishment (measured as length of sentence in years), age at the time of arrest, presence of a prior criminal record (yes or no), and ascribed socioeconomic class (middle, working, and low) was collected for 37 convicted robbery offenders. These hypothetical data are shown below:

Offender ID	Length of Sentence	Age	Prior Record	Socioeconomic Class
1	20	21	Yes	Working
2	10	24	No	Low
3	4	34	No	Low
4	13	45	Yes	Working
5	20	24	Yes	Working
6	22	30	No	Working
7	15	32	Yes	Low
8	6	45	Yes	Low
9	18	28	Yes	Working
10	20	23	Yes	Low
11	16	23	No	Working
12	25	22	Yes	Low
13	10	35	No	Low
14	14	22	No	Middle
15	9	24	No	Middle
16	20	38	Yes	Middle
17	4	36	No	Middle
18	18	40	Yes	Middle
19	14	22	Yes	Middle
20	12	23	Yes	Middle
21	9	25	Yes	Working
22	17	24	No	Working
23	8	48	Yes	Working
24	6	49	Yes	Low

Offender ID	Length of Sentence	Age	Prior Record	Socioeconomic Class
25	20	22	No	Working
26	20	45	Yes	Working
27	22	23	Yes	Working
28	24	31	Yes	Working
29	7	24	No	Low
30	3	22	No	Middle
31	4	23	No	Low
32	10	24	No	Working
33	1	21	No	Middle
34	9	25	No	Low
35	20	22	Yes	Working
36	22	23	Yes	Working
37	6	27	No	Middle

1. From the above data, use multiple regression to test the research hypothesis that age, prior criminal record, and ascribed socioeconomic class influence severity of punishment.
2. From the multiple regression in Question 1, predict the sentence length for a 65-year-old working class offender with a prior record. Should you trust this predicted value? Explain.

3.7.2 Exercise 2: Fire Department Size

A researcher is using data from 106 cities to study fire department size. Specifically, she is interested in using a regression analysis to predict the number of firefighters per 10,000 population. The regression below used the independent variables percentage unemployed, percentage of adults holding a bachelor's degree, and region (NE, MW, S, or W). In these (hypothetical) data, the firefighter variable ranged from 11.28 to 24.81 (mean = 17.3, standard deviation = 3.4). Percentage unemployed ranged from 7.1% to 11.6% (mean = 9.3, s.d. = 1.1), and percentage with a bachelor's degree ranged from 16.3% to 28.7% (mean = 22.9, s.d. = 2.9) across the 106 cities.

Her regression results were as follows:

Dependent variable: number of firefighters per 10,000 population

N = 106
R^2 = 0.534
Sum of squared errors = 501.82

	b	SE of b	t	p-value
Intercept (constant)	8.56	1.34		
% unemployed	−1.02	0.37	(i)	0.007
% bachelor's degree	0.74	0.54	1.37	0.173
NE dummy	2.13	0.83	2.57	0.011
MW dummy	1.52	0.68	2.24	0.027
W dummy	−0.87	0.59	−1.47	0.144

In a regression with only % unemployed and % bachelor's degree as independent variables, the sum of squared errors was 553.29.

1. What is the value for the omitted number (i) in the output table?
2. Use the results to predict the number of firefighters per 10,000 population for a city with 8% unemployment, 22% holding a bachelor's degree, and located in the Midwest. Should you trust this predicted value? Explain.
3. If the analyst had chosen "W" as reference, what would be the b's for "NE," "MW," and "S"?
4. If "W" had been chosen as reference, what would happen to R^2, b for % unemployed, and b for % holding bachelor's degree?
5. Assuming no collinearity issue, interpret the regression results as fully as possible.

3.7.3 Exercise 3

1. State the two competing hypotheses involved in assessing the impact of a set of multiple (more than one) dummy variables on the dependent variable. How do you determine whether to reject or not reject the null hypothesis?
2. A researcher used survey data in which respondents were categorized by political identification as "Republican," "Democratic," or "other." For his regression analysis, the researcher created a political ID variable by haphazardly assigning numbers to each category (such as Republican = 1, Democratic = 2, and Other = 3), and included this as a numerical independent variable. What assumptions does this approach make?

3.8 Appendix

3.8.1 The General Form of an F-Test

The applications of the F-test shown in this chapter involved the assessment of statistical significance for sets of dummy variables. The F-test, however, actually has much wider applicability, and the uses of it in this chapter are simply special cases of a more general setup. The general form of an F-test is as a comparison of two "nested" regression models. Two models (for the same dependent variable Y) are nested if one model includes all the variables in the other, plus some more.

For example, if the independent variables in one model predicting income are age, years of education, and a gender dummy, and the independent variables in a second model predicting income are age, years of education, a gender dummy, and a set of dummies for race/ethnicity, then the two models are nested. The second model includes all the independent variables in the first, plus the additional set of race/ethnicity dummy variables. On the other hand, if the second model had included age, a gender dummy, and the set of race/ethnicity dummies, the two models would not be nested. That is because the second model would include some variables not in the first (the race/ethnicity dummies), but not all of the variables that were in the first (because the education variable is absent). In the examples of the F-test discussed throughout this chapter, the two models being compared were nested because one model (Model 1) added a set of dummies while retaining all of the variables in the other (Model 0).

The formula shown in the chapter for calculating F in tests of a set of dummy variables is valid for comparing any two nested regression models, with the only change being that s in the F-statistics formula will more generally represent the number of additional variables that are included in the model that has more variables. Beyond the dummy variable application already discussed, why might we use the F-test to compare two *nested models?* One situation in which the F-test is useful is when the set of independent variables can be divided into one set representing the variables associated with a specific theory and a second set of "control" variables (which we discussed earlier as being included in the regression because they are thought to be important in predicting Y, despite not being the focus of the theory that the researcher is investigating).

Then an F-test could be used to test whether the variables indicated by the theory significantly improve on the model that includes only the control variables. Using the notation of this chapter, Model 0 would include the control variables only, while Model 1 would include the control variables plus the set of theoretically inspired variables. Another situation for the F-test could involve two theories that are nested in the sense that all of the independent variables suggested by theory A are also suggested by theory B, but theory B adds some additional ones. Then Model 0 would include the independent variables common to the two theories, and Model 1 would add the variables unique to theory B.

Another scenario in which the F-test can be useful is as a response to collinearity. As discussed in the previous chapter, collinearity makes it difficult to distinguish which X's are or are not having effects on Y. There we addressed collinearity by dropping X's that were highly correlated with other X's in the model, but we were careful to note that the decisions of which X's to drop were essentially arbitrary. In pairs of X's from which we dropped a variable, we could not definitively decide which (or if both) of the X's in the pair were actually having an impact on Y. We can take this one step further and focus directly on a set of highly correlated X's, not the individual X's in the set. This can be done via the F-test. Model 1 would include among its X's the set of highly correlated X's, and this set would be removed for Model 0. Then the F-test indicates whether the set of highly correlated X's together have a significant effect on Y. If the F-test finds that these X's do significantly impact Y, we still may not be able to determine which of these X's are or are not having an effect on Y, but the results indicate that at least some do.

3.8.2 The F Shown by Default on Regression Output

We mentioned that the regression output from any standard statistical software will always show a specific F-statistic and its associated p-value, and we usually do not use this F for an assessment of statistical significance for a set of dummy variables. What two nested models are being compared with this default F in the output? This F-statistic compares our regression model, with whatever independent variables we chose to include, to the "naïve," or "no independent variable" model in which we predict the same value of Y for each case in our sample. In the *naïve model* we do not use the X's to make different predictions of Y for different cases in the sample but instead make the same prediction of Y for every person or thing in the sample. In our regression language, the naïve model is simply $\hat{Y} = a$, and the estimate a will turn out to be the sample mean of Y. That is because when we have to make the same prediction of Y for every case in our sample, the sample mean ($\overline{Y}$) will be the best choice for that prediction. The naïve model and our full regression model (assuming that it includes an intercept, as is the case for all of the regressions discussed in this book) are nested: the naïve model has no independent variables, so we could say, albeit somewhat awkwardly, that any independent variable in the naïve model also appears in the full regression model.

We are usually not too interested in this comparison to the naïve model. If our regression model reflects a reasonable theory, past research, or even just common sense in choosing X's, it is very likely that our model is a good deal better at predicting Y than the intercept-only naïve model is. Viewing each case as identical to all the others, so that every case will have the same predicted value of Y, is almost surely too simplistic. For a realistic regression model, then, we expect that the improvement in fit of our regression over the naïve model will be statistically significant, so the p-value associated with this F-statistic will almost always be very small. In this respect the F-test shown on the regression output is typically not very interesting or informative. Of course if in some data set this F-test yielded a large p-value, we would certainly notice this and understand that our complicated model's predictions were not enough better than those of the naïve model to be worth the additional df used by the more complicated model. Again, though, this will be rare if we have chosen our X's sensibly.

However, one exception to our usual lack of interest in this default F-statistic concerns regressions in which the only independent variables are a set of dummies representing a categorical independent variable. This type of regression turns out to be closely related to a seemingly distinct use of the F-statistic. We describe this connection in Section 3.8.3.

3.8.3 F and ANOVA

Some readers may have experience with *analysis of variance,* also known as *ANOVA*. The classic ANOVA problem is to test the null hypothesis that the mean of some variable Y is the same in each of several group populations; this is most straightforward when there are equal-sized samples from each group. For example, we may be interested in testing the null hypothesis (H_0) that mean income is the same in the populations of four different racial/ethnic groups, against the alternative hypothesis (H_1) that there is at least some inequality (difference) in mean income across the groups. (That is, the alternative hypothesis holds that it is not the case that the means are all equal but does not specify how they depart from equality.) In testing this via samples of equal size from the various groups, an F-statistic is used to generate a p-value for the null hypothesis of equal means across the group populations.

This classic ANOVA problem can be reformulated as a multiple regression with dummy variables indicating membership in the groups. In the example above, we can analyze the sample data by treating income as Y and creating three dummy variables to represent the four-category race/ethnicity variable. This regression has no other independent variables but the race/ethnicity dummy variables, with the intercept of course there too. Certainly we can immediately think of other independent variables that we would also want to include as predictors of income, and that really should be controlled for when examining racial/ethnic differences in income. But suppose for this example that we had no such variables at hand because the survey asked respondents for only their race/ethnicity and their income.

To test the statistical significance of race/ethnicity using the methods discussed in this chapter, we would use the F-statistic to compare the model with the set of race/ethnicity dummy variables (Model 1) to the model that omits that set (Model 0). Because there are no other independent variables, Model 0 is simply the intercept and nothing else—the naïve model. In that case, the default F shown on the output for the regression with the set of race/ethnicity dummy variables (Model 1) is precisely what we want in order to test the statistical significance of race/ethnicity. Furthermore, the F-statistic shown there would equal the F-statistic that we would calculate in a different way if we followed the ANOVA approach in this classical setting.

Regression with dummy variables therefore has a close connection to classical ANOVA. There are other areas of overlap with the broader ANOVA framework, including with regression models that contain multiple categorical variables. The ANOVA framework has been intensely studied and extended over many decades in several academic fields and is now very rich. Extensions of the classical setup can become very complicated, so we do not explore any further here, but advanced texts can be consulted for more information.

4

Multiple Regression With Interaction

In Chapter 3, we learned how to include categorical independent variables in multiple regression. This chapter introduces another extension of multiple regression analysis, known as *interaction*, incorporating the idea that the effect of an X on Y may vary depending on the value of another X. This chapter begins by explaining the general idea of interaction then shows how interaction terms are used to implement this idea in multiple regression. We first examine the special case of an interaction between a categorical and a numerical independent variable, followed by a full example. We then discuss additional situations of interaction between two numerical variables and between two categorical variables, concluding with exercises.

4.1 Interaction

When first hearing the word "interaction," we might think that simply including two X's in a regression means that there is an interaction between them, because it seems that together they are "interacting" to produce a predicted value of Y. That is **not** what we mean by an interaction between two X's. Instead, an interaction between two X's means that each variable's effect on Y depends on the value (or category) of the other variable. That is, if there is an interaction between X_1 and X_2, then the effect of X_1 on Y will be different when $X_2 = 10$ than it is when $X_2 = 20$. Likewise, the effect of X_2 will be different depending on what value X_1 takes. Interaction is sometimes referred to as *"moderation,"* with X_1 moderating the effect of X_2 on Y (and vice versa).

For instance, it may be that the effect of a person's education (X_1) on their income (Y) differs depending on the person's ascribed socioeconomic class, possibly measured as income of the person's household when they were 15 years old (X_2). Perhaps an additional year of education has a greater effect on income for individuals whose parents had high incomes than for those whose parents had low incomes. Or it may be that the effect of education on income differs in different regions of the country; if so, this would be an interaction between a person's education and region of residence (with region represented by a set of dummy variables). In research on high school students' satisfaction (Y) with school, we may suspect that the predicted difference in satisfaction between members and nonmembers of athletic teams is not the same for those who participate in some nonathletic school group and those who do not. This would be an interaction between athletic team membership and nonathletic group participation. As suggested by the above examples, interaction is a general idea that can involve independent variables of any type. It is possible to have an interaction between a categorical and a numerical independent variable, between two numerical variables, or between two categorical variables. The idea of interaction actually even generalizes to a situation involving three (or even more) variables. Such multiway interactions must be interpreted very carefully, and they are less often encountered in practice, so we restrict our attention here to interactions between two variables.

4.2 Interaction Between a Categorical and a Numerical Independent Variable

In introducing the details of interaction and its interpretation, we will first focus on the special case of an interaction between a categorical and a numerical independent variable. The additional situations of interaction between two numerical variables and between two categorical variables involve the same logic and will be discussed later in the chapter.

4.2.1 Construction of Interaction Terms

To build the notion of interaction into a multiple regression analysis, specially constructed variables need to be included as additional independent variables. Before showing how these additional variables are constructed, it is helpful to first revisit the presentation of multiple regression with dummy variables that we introduced in Chapter 3. This will make it easier to grasp how the interaction terms work to operationalize the interaction concept.

Suppose we have a regression model in which the dependent variable Y is predicted by a numerical independent variable X_1 as well as the categorical independent variable "class in school" (X_2). The class in school variable (X_2) has four categories: "Freshman," "Sophomore," "Junior," and "Senior." The regression model may be written as

$$\hat{Y} = a + b_1X_1 + b_2X_{2a} + b_3X_{2b} + b_4X_{2c}$$

where X_{2a} is a dummy variable for "Freshman," X_{2b} is a dummy variable for "Sophomore," and X_{2c} is a dummy variable for "Junior," with "Senior" left out as the reference category. This model has no interaction and will be our starting point.

To understand how an interaction between school class and X_1 works, it is helpful to first write out this starting model as a separate equation for each class in school, as we did when first presenting dummy variables in Chapter 3. For freshmen, the "Freshman" dummy X_{2a} will equal 1, and the other dummies will be zero. So for freshmen, the regression equation is

$$\begin{aligned}\hat{Y} &= a + b_1X_1 + b_2X_{2a} + b_3X_{2b} + b_4X_{2c}\\ &= a + b_1X_1 + b_2(1) + b_3(0) + b_4(0)\\ &= a + b_1X_1 + b_2,\end{aligned}$$

which can be rewritten as

$$\text{Freshmen: } \hat{Y} = (a + b_2) + b_1X_1.$$

In the same way, the equations for the other classes, remembering that "Senior" is the reference category and has zero for every dummy, are

$$\begin{aligned}\text{Sophomores: } \hat{Y} &= a + b_1X_1 + b_2(0) + b_3(1) + b_4(0)\\ &= a + b_1X_1 + b_3 = (a + b_3) + b_1X_1.\\ \text{Juniors: } \hat{Y} &= a + b_1X_1 + b_2(0) + b_3(0) + b_4(1)\\ &= a + b_1X_1 + b_4 = (a + b_4) + b_1X_1.\\ \text{Seniors: } \hat{Y} &= a + b_1X_1 + b_2(0) + b_3(0) + b_4(0)\\ &= a + b_1X_1.\end{aligned}$$

By grouping the terms in parentheses this way, we can see that the dummy variables are producing different intercepts—$(a + b_2)$ for freshmen, $(a + b_3)$ for sophomores, $(a + b_4)$ for juniors, and simply a for seniors—for the regression equations that apply to the different classes. Notice, though, that the effect of X_1 (its slope) is the same, b_1, for every class. This usual setup of dummy variables, without interaction, therefore creates a situation in which the intercept of the regression line is different in the different classes, but the slope with respect to X_1 is the same.

Including an interaction between class and X_1 allows the slopes of the regression lines to be different in the different classes, because it permits the effect of X_1 to change depending on the value (in this case, the category) of the variable school class. In other words, an interaction allows the effect of X_1 on Y to be different for freshmen, sophomores, juniors, and seniors. We can build this into the regression by adding specially constructed new variables that are called interaction variables or *interaction terms* (or moderator variables, if we were referring to interaction as moderation). These are created by multiplying X_1 with each of the dummy variables associated with the categorical variable school class. School class has four categories, and therefore three dummy variables (X_{2a}, X_{2b}, and X_{2c}), so the interaction with X_1 will require three new interaction terms, calculated as $X_1 \times X_{2a}$, $X_1 \times X_{2b}$, and $X_1 \times X_{2c}$.

For example, suppose the classes (and the dummy variables for class) and values of X_1 for the first five respondents in the data were as follows:

Person ID	Class	X_1	Dummy Variables for School Class		
			Freshman Dummy (X_{2a})	Sophomore Dummy (X_{2b})	Junior Dummy (X_{2c})
1	Sophomore	15	0	1	0
2	Freshman	22	1	0	0
3	Senior	30	0	0	0
4	Junior	27	0	0	1
5	Freshman	20	1	0	0

The resulting new interaction terms have been added to this table:

Person ID	Class	X_1	Dummy Variables for School Class			Interaction Terms or Variables		
			Fresh. (X_{2a})	Soph. (X_{2b})	Junior (X_{2c})	$X_1 \times X_{2a}$ (X_1X_{2a})	$X_1 \times X_{2b}$ (X_1X_{2b})	$X_1 \times X_{2c}$ (X_1X_{2c})
1	Sophomore	15	0	1	0	0	15	0
2	Freshman	22	1	0	0	22	0	0
3	Senior	30	0	0	0	0	0	0
4	Junior	27	0	0	1	0	0	27
5	Freshman	20	1	0	0	20	0	0

If a dummy variable had value 0 for a respondent, the corresponding new variable will also have value 0. If a dummy variable had value 1 for a respondent, the corresponding new variable will take whatever value X_1 had for that respondent. To represent interaction, these new variables will be included in the regression along with X_1 and the original dummy variables for class. With the interaction terms, the regression equation is

$$\hat{Y} = a + b_1X_1 + b_2X_{2a} + b_3X_{2b} + b_4X_{2c} + b_5X_1X_{2a} + b_6X_1X_{2b} + b_7X_1X_{2c}.$$

4.2.2 Interpretation of b's for Interaction Terms

As before, it is helpful to write the regression equation separately for each class, again grouping terms when possible:

$$\hat{Y} = a + b_1X_1 + b_2X_{2a} + b_3X_{2b} + b_4X_{2c} + b_5X_1X_{2a} + b_6X_1X_{2b} + b_7X_1X_{2c}.$$

$$
\begin{aligned}
\text{Freshmen:} \quad \hat{Y} &= a + b_1X_1 + b_2(1) + b_3(0) + b_4(0) + b_5X_1(1) \\
&\quad + b_6X_1(0) + b_7X_1(0) \\
&= a + b_1X_1 + b_2 + b_5X_1 = (a + b_2) + (b_1 + b_5)X_1. \\
\text{Sophomores:} \quad \hat{Y} &= a + b_1X_1 + b_2(0) + b_3(1) + b_4(0) + b_5X_1(0) \\
&\quad + b_6X_1(1) + b_7X_1(0) \\
&= a + b_1X_1 + b_3 + b_6X_1 = (a + b_3) + (b_1 + b_6)X_1. \\
\text{Juniors:} \quad \hat{Y} &= a + b_1X_1 + b_2(0) + b_3(0) + b_4(1) + b_5X_1(0) \\
&\quad + b_6X_1(0) + b_7X_1(1) \\
&= a + b_1X_1 + b_4 + b_7X_1 = (a + b_4) + (b_1 + b_7)X_1. \\
\text{Seniors:} \quad \hat{Y} &= a + b_1X_1 + b_2(0) + b_3(0) + b_4(0) + b_5X_1(0) \\
&\quad + b_6X_1(0) + b_7X_1(0) \\
&= a + b_1X_1.
\end{aligned}
$$

It is easy to see that the interaction terms have done what we wanted: the effect of X_1 on $\hat{Y}$ now changes depending on the class in school. The estimated effect of X_1 on Y is $(b_1 + b_5)$ for freshmen, $(b_1 + b_6)$ for sophomores, $(b_1 + b_7)$ for juniors, and just b_1 for seniors. That is, for freshmen, the effect of X_1 on Y comes from adding the b for the interaction between X_1 and the Freshman dummy (b_5) to the b for X_1 alone (b_1). For sophomores, this effect is the b for X_1 alone (b_1) plus the b for interaction between X_1 and the Sophomore dummy (b_6), and for juniors the effect is the b for X_1 alone (b_1) plus the b for the interaction between X_1 and the Junior dummy (b_7). For the reference category seniors, the effect of X_1 is simply given by the b for X_1 alone (b_1).

Suppose that the regression output showed $b_1 = 8.5$, $b_5 = 2.2$, $b_6 = -1.6$, and $b_7 = -0.4$. Then the estimated effect of X_1 on $\hat{Y}$ would be $b_1 + b_5 = 8.5 + 2.2 = 10.7$ for freshmen, $b_1 + b_6 = 8.5 + (-1.6) = 6.9$ for sophomores, $b_1 + b_7 = 8.5 + (-0.4) = 8.1$ for juniors, and $b_1 = 8.5$ for seniors. From these values, the estimated effect of X_1 on Y is greatest for freshmen and least for sophomores. The equations above also show why b_5 is interpreted as the difference in the (estimated) effect of X_1 on Y between freshmen and seniors, b_6 gives the difference in the effect of X_1 on Y between sophomores and seniors, and b_7 indicates the difference in the effect of X_1 on Y between juniors and seniors.

This discussion shows how including the interaction implies that we are estimating a different effect of X_1 for each of the school classes. One consequence of this is that it will be desirable to be sure that our sample includes an adequate number of cases in each class, rather than simply focusing on the overall sample size. We are being more demanding of our data when we ask the regression to estimate X_1's effect within each class, instead of simply estimating a common effect of X_1 that applies to the whole sample. Another fact becomes relevant when we envision a regression model that includes school class, X_1, and additional independent variables, say X_3 and X_4. If were to set up a further regression that included terms for interactions between school class and *all* of the other independent variables, X_1, X_3, and X_4, the resulting analysis would in effect be conducting a separate regression with independent variables X_1, X_3, and X_4 within each of the school classes. In practice we tend to be reluctant to include very many interactions in a regression, but sometimes the idea of a completely distinct regression within each group will be consistent with a social science theory that we are investigating.

The interpretation so far has focused on the way that the effect of X_1 differs across classes. The interaction can also be interpreted in the opposite way, focusing on how the differences between $\hat{Y}$ for the different classes depend on the value of X_1. To see this, it is helpful to rewrite the regression equation in still one more way, again by grouping terms:

$$\hat{Y} = a + b_1X_1 + b_2X_{2a} + b_3X_{2b} + b_4X_{2c} + b_5X_1X_{2a} + b_6X_1X_{2b} + b_7X_1X_{2c}$$
$$= a + b_1X_1 + (b_2b_5X_1)X_{2a} + (b_3 + b_6X_1)X_{2b} + (b_4 + b_7X_1)X_{2c}.$$

In this rewritten equation, $(b_2 + b_5 X_1)$ is associated with the Freshman dummy variable (X_{2a}) as the difference in $\hat{Y}$ between freshmen and seniors. Similarly, $(b_3 + b_6 X_1)$ is associated with the Sophomore dummy variable (X_{2b}), as the difference in $\hat{Y}$ between sophomores and seniors, and $(b_4 + b_7 X_1)$, for the Junior dummy (X_{2c}), is the difference in $\hat{Y}$ between juniors and seniors. Differences in $\hat{Y}$ between these three categories and the reference category senior (as well as differences among the three categories) therefore change depending on the value of X_1 that is plugged into these expressions.

For illustration, let us examine the coefficient of the Freshman dummy (X_{2a}), showing the difference in $\hat{Y}$ between freshmen and seniors. From the above equation for the model with interaction, the coefficient on the Freshman dummy (X_{2a}) on Y is $(b_2 + b_5 X_1)$. Suppose that $b_2 = 11.6$, and $b_5 = 2.2$ as before. Also suppose that we are interested in finding the difference in $\hat{Y}$ between freshmen and seniors when $X_1 = 24$. From the equation, the difference in $\hat{Y}$ between freshmen and seniors is $(b_2 + b_5 X_1) = 11.6 + (2.2 \times 24) = 11.6 + 52.8 = 64.4$. Of course this difference will change if we look at a different value of X_1. When $X_1 = 50$, this figure is $(b_2 + b_5 X_1) = 11.6 + (2.2 \times 50) = 11.6 + 110 = 121.6$. In this case, we observe that the difference in the predicted value of Y between freshmen and seniors grows larger at higher levels of X_1, because b_5 is positive. Both kinds of interpretation—(a) the difference in the effect of X_1 on Y in the different classes and (b) how the differences in $\hat{Y}$ between classes depend on the value of X_1—are inherent in the interaction. Often one interpretation will seem more relevant to the research questions that we are investigating, and if so we will focus on that, but both are always present, as two sides of the same coin.

4.2.3 Statistical Significance of Interaction Terms

Of course one of our main concerns is whether the interaction between X_1 and class is statistically significant. If not, then the additional complexity that the interaction terms introduce to the regression equation is not justified, and we can base our interpretations on the simpler model that does not include interaction. The assessment of statistical significance for an interaction parallels the assessment of statistical significance in a set of dummy variables: if the interaction adds only one newly constructed interaction term/variable to the model, we can simply use the usual t-test for that one new term, but if it adds more, we will need to assess statistical significance of the interaction via an F-test. This F-test will compare the model that includes the interaction (Model 1) to the model without the interaction (Model 0). In this case, s, the degrees of freedom for the numerator, will be the number of new interaction terms. One slight wrinkle is that it is possible for the interaction to be statistically significant even if the underlying set of dummy variables or the numerical X_1 are not. In the above example, for instance, if b_1 were not statistically significant, it would mean that there was no statistically significant effect of X_1 for seniors (see the regression equation for seniors above). But there could be statistically significant effects of X_1 for the other classes, which would be captured by the interaction terms. In such a case we will still proceed with interpretation of the interaction, typically still including the b's for the dummy variables and X_1 in our interpretations as above.

In discussing the F-test for a set of dummy variables, we noted that the F-test could indicate statistical significance for the categorical variable overall even if some of the p-values from t-tests of each dummy's coefficient were large. The same thing can happen with interactions: the F-test could indicate statistical significance of the interaction as a whole while some of the specific interaction terms do not show statistical significance. So, as with a set of dummies, the overall statistical significance of the interaction does not necessarily mean that all of the differences in the effect of X_1 across categories are themselves statistically significant (or, alternatively, that all of the between-category differences significantly vary according to the value of X_1). Similarly, a small p-value for one or more of the interaction terms does not necessarily mean that the F-test will indicate statistical significance for the set.

If our regression has several categorical independent variables, requiring several sets of dummy variables, and several numerical independent variables, there will be many possible interactions. Because interactions make the model's results more difficult to interpret, typically analysts investigate only a small number of the possible interactions, or perhaps even none at all. The complicated interpretation involved in an interaction means that we will generally only examine those possible interactions that are directly suggested by theory or seem important in a preliminary examination of our data.

4.3 Example: Income, Education, and Work Experience in the Legal World

Here is an example in which the researcher's working hypotheses suggest that there should be an interaction between two independent variables. The example is an illustration of interaction between a categorical independent variable and a numerical independent variable. This illustration will help familiarize us with how interaction terms are created and included in the regression analysis and how results are interpreted.

A researcher was interested in income among people holding various kinds of jobs within the field of law. He hypothesized that greater work experience is associated with higher income. Furthermore, he believed that there are differences in average income among holders of different academic degrees and that the effect of work experience on income differs for people with different degrees; the last part of this statement suggests that there is an interaction between work experience and degree. He selected a sample of 27 people working in occupations related to the practice of law and used their data to test his hypotheses with multiple regression analysis. (The data below are hypothetical, not from a real sample; note that this is also a small sample in which to examine interaction, but the small sample size allows us to provide the full data here for any reader who wishes to reproduce the results.) Work experience was measured for each person by years in the labor force. Academic degrees included three categories: "PhD," "JD," and "BA," and income was measured by annual income (in thousands).

The research hypothesis, units of analysis, measurements of independent and dependent variables, and raw data are as follows:

Research hypothesis: People with greater work experience will have greater income. There are differences in income among holders of different academic degrees. Also, the effect of work experience on income differs for people depending on their academic degree. An alternative to the last statement is that the differences in income among holders of different academic degrees depend on the level of work experience. (Either of these two ways of saying this indicates an interaction between experience and degree; the researcher's choice of which to explicitly state will probably depend on "which side of the coin" will seem more interesting to his intended audience.)

Units of analysis: 27 workers in the field of law (individual-level data).

Measurements of independent and dependent variables:

Y: Annual income in thousands (income).

X_1: Years in labor force (work experience).

X_2: Academic degree: "PhD," "JD," and "BA" (education).

Research hypothesis with independent and dependent variable measures:

1. Years in labor force (X_1) have a positive effect on annual income (Y).
2. There are differences in annual income (Y) among people with different academic degrees (X_2).
3. The effect of years in labor force (X_1) on annual income (Y) changes depending on a person's type of degree (X_2). Alternatively, differences in annual income (Y) between holders of different academic degrees (X_2) depend on their years in the labor force (X_1).

Data:

Worker ID	Income (Y)	Years in Labor Force (X_1)	Academic Degree (X_2)
1	20	2	JD
2	43	8	BA
3	43	5	JD
4	72	9	PhD
5	76	11	JD
6	37	10	BA
7	34	4	PhD
8	59	9	JD
9	27	4	PhD
10	56	8	JD
11	62	10	JD
12	84	12	JD
13	30	5	PhD
14	25	3	BA
15	63	12	BA
16	49	9	BA
17	71	10	JD
18	77	8	JD
19	16	1	PhD
20	48	8	BA
21	88	11	JD
22	15	4	PhD
23	39	6	BA
24	70	10	PhD
25	40	1	BA
26	48	4	JD
27	51	3	PhD

Data source: Hypothetical data.

Because academic degree (X_2) is a categorical variable, it must be represented through dummy variables. There are three categories of academic degree ("PhD," "JD," and "BA"), so two dummy variables must be created. Any category could be the reference; suppose that the researcher selected "BA." He would then need to create dummy variables for the remaining categories "PhD" (X_{2a}) and "JD" (X_{2b}). These dummy variables would look as follows:

Worker ID	Academic Degree (X_2)	Dummy Variables for Academic Degree	
		"PhD" Dummy (X_{2a})	"JD" Dummy (X_{2b})
1	JD	0	1
2	BA	0	0
3	JD	0	1
4	PhD	1	0
5	JD	0	1
6	BA	0	0
7	PhD	1	0
8	JD	0	1
9	PhD	1	0
10	JD	0	1
11	JD	0	1
12	JD	0	1
13	PhD	1	0
14	BA	0	0
15	BA	0	0
16	BA	0	0
17	JD	0	1
18	JD	0	1
19	PhD	1	0
20	BA	0	0
21	JD	0	1
22	PhD	1	0
23	BA	0	0
24	PhD	1	0
25	BA	0	0
26	JD	0	1
27	PhD	1	0

After entering the above data in statistical software, he ran a correlation analysis (realizing, as above, that the correlations between dummies from the same categorical independent variable will often be fairly large). The results of the correlation analysis are below.

	Years in Labor Force (X_1)	"PhD" Dummy (X_{2a})	"JD" Dummy (X_{2b})
Years in labor force (X_1)	1.000	−0.370	0.308
"PhD" dummy (X_{2a})	−0.370	1.000	−0.538
"JD" dummy (X_{2b})	0.308	−0.538	1.000

The highest correlation was between the "PhD" (X_{2a}) and "JD" (X_{2b}) dummy variables ($r = -0.538$), but the numerical variable years in labor force (X_1) was not too highly correlated with either dummy. As a rather strong negative correlation between the two education dummies is not surprising, these correlations do not indicate a potential problem of collinearity.

The research hypothesis suggests an interaction between years in labor force (X_1) and academic degree (X_2), but it is still a good idea to begin the analysis with a regression model that does not include interaction. Results from a multiple regression using years in labor force (X_1), the "PhD" dummy variable (X_{2a}), and the "JD" dummy variable (X_{2b}) as the independent variables are shown below.

$R^2 = 0.726$
df for t = 23

	b	SE of b	t	p-value
Intercept (constant)	10.819	6.574	1.646	0.113
Years in labor force (X_1)	4.517	0.720	6.272	< 0.001
"PhD" dummy (X_{2a})	5.973	6.009	0.994	0.331
"JD" dummy (X_{2b})	14.409	5.454	2.642	0.015

The effect of years in labor force (X_1) on annual income (Y) is statistically significant (as indicated by a p-value of < 0.001, which may be shown as "0.000" in software output), controlling for academic degree (X_2). For each additional year in the labor force (X_1), predicted annual income ($\hat{Y}$) increases 4.517 (thousands), controlling for academic degree (X_2). That is, holding academic degree (X_2) constant, predicted annual income ($\hat{Y}$) increases by $4,517 for each additional year in the labor force. The sample averages 6.93 years in the labor force, so 1 additional year of experience seems like a reasonable change to consider, and the corresponding estimated increase of $4,517 in predicted income seems like a large enough effect to be considered real-world significant, given the mean ($49,740) and the range ($15,000–$88,000) of annual income (Y).

To test the statistical significance of the set of dummy variables for academic degree (X_2), an F-statistic comparing the regression models with and without the set of degree dummies must be calculated. The sum of squared errors for the regression above (Model 1) was 3,107.067, and for another regression (not shown) without the degree dummies (leaving only X_1; Model 0) the sum of squared errors was 4,062.997. With a set of 2 dummies being tested, and 23 df for the full regression, the F calculation is

$$\begin{aligned} F &= \frac{(SSE_0 - SSE_1)/s}{SSE_1/df_1} = \frac{(4,062.997 - 3,107.067)/2}{3,107.067/23} \\ &= \frac{955.93/2}{3,107.067/23} = \frac{477.965}{135.090} = 3.538. \end{aligned}$$

With (2, 23) df, the F-table (page 237) shows that this F-statistic yields a p-value below the 0.05 cutoff (between 0.025 and 0.05), so there is a statistically significant impact of the academic degree variable (X_2) on income. (A more precise p-value is 0.046, from statistical software or an online calculator.) Interpreting the b's associated with the dummy variables, the estimated difference between people with a PhD (X_{2a}) and those with just a BA (the reference), controlling for years in labor force (X_1), is 5.973 thousands of dollars of predicted annual income. That is, the difference in predicted annual income is $5,973 between people with a PhD and those with just a BA, controlling for years in labor force (X_1), with people holding a PhD having a higher predicted income than those holding a BA only. Note, however, that the p-value for the PhD dummy is large ($p = 0.331$), so this specific comparison of these two educational levels does not indicate a statistically significant difference in predicted income. As the overall assessment (via the F-test) of the education variable indicated statistical significance, there must be some other statistically significant difference among the categories.

The difference in predicted annual income in thousands ($\hat{Y}$) between people with a JD (X_{2b}) and those with just a BA (the reference) is 14.409, and statistically significant ($p = 0.015$), controlling for years in labor force (X_1). In dollars rather than thousands of dollars, this difference in predicted annual income ($\hat{Y}$) means that people with a JD are predicted to earn $14,409 more than those with just a BA (the reference category), holding years in labor force (X_1) constant.

For a graphical representation of the differences in predicted income among the educational categories, we can place the dummies' b's and the reference category's 0 on the number line. We can also indicate the spacing between adjacent categories:

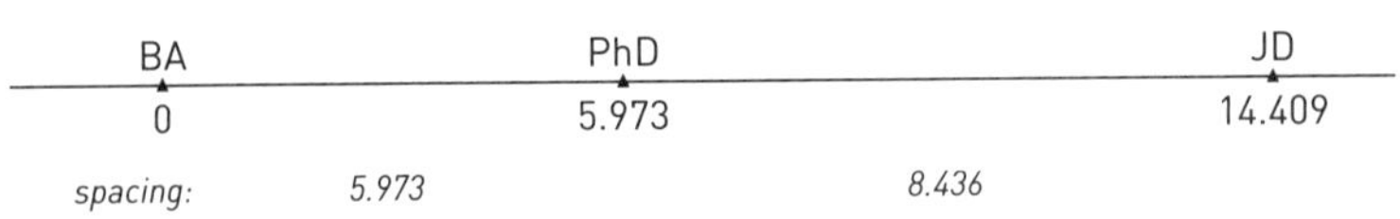

From the number line above, it is easy to see that predicted income (controlling for years in labor force X_1) is highest for people with a JD, followed by people with a PhD, and finally by those whose highest degree is BA. Also from the number line, the difference in predicted income (Y) between dummy categories (PhD and JD) is \$8,436 (from 14.409 – 5.973). To examine the statistical significance of the difference between PhD and JD, we need to switch the reference category to either PhD or JD, and rerun the model. When we do this (not shown), the difference is not statistically significant ($p = 0.164$). In sum, the F-test indicated that, considered as a whole variable, academic degree (X_2) had a statistically significant relationship with income (Y), but the only statistically significant specific difference behind this overall assessment was that between BA and JD. Taking into account the mean (\$49,740) and the range (\$15,000–\$88,000) of annual income (Y), the difference of \$14,409 in predicted income (Y) between the BA and JD categories seems large enough to be considered important in real terms.

R^2 is 0.726. This indicates that the regression model is performing rather well in explaining or predicting annual income (Y), with 72.6% of the variance in income (Y) explained by years in labor force (X_1) and academic degree (X_2). Together, these X's seem to do a pretty good job of accounting for differences in income (Y) in this sample. There may be other X's that could be added to make R^2 even higher, but this model seems to have successfully captured important influences on income (Y). With this relatively high R^2, and with both X's statistically significant (and, for the numerical variable work experience, in the expected positive direction), the first two parts of the research hypotheses—that years in labor force (X_1) has a positive effect on annual income (Y) and that there are differences in annual income (Y) among people with different academic degrees (X_2)—seem to be supported by this analysis. There is a caveat, however, as not all of the estimated income differences between specific degrees were statistically significant.

Next the researcher turned to the third part of the research hypothesis, considering whether the effect of work experience (X_1) on income (Y) changes depending on a person's type of degree (X_2). Expressed in the opposite terms, he is asking whether the difference in income (Y) among holders of different academic degrees depends on the level of work experience (X_1). Either way of expressing the hypothesis indicates an interaction between degree and work experience, so testing this hypothesis requires the addition of interaction terms/variables for an interaction between years in labor force (X_1) and academic degree (X_2). Two dummy variables represent academic degree (X_2), so the researcher needed to create two new variables to represent this interaction. One is created by multiplying X_1 with the "PhD" dummy (X_{2a}), and the other by multiplying X_1 with the "JD" dummy (X_{2b}). The data with the interaction terms look like this:

		Academic Degree (X_2)		Interaction Between X_1 and X_2	
Worker ID	Years in Labor Force (X_1)	"PhD" (X_{2a})	"JD" (X_{2b})	$X_1 \times X_{2a}$ (X_1X_{2a})	$X_1 \times X_{2b}$ (X_1X_{2b})
1	2	0	1	0	2
2	8	0	0	0	0
3	5	0	1	0	5

(Continued)

Worker ID	Years in Labor Force (X_1)	Academic Degree (X_2)		Interaction Between X_1 and X_2	
		"PhD" (X_{2a})	"JD" (X_{2b})	$X_1 \times X_{2a}$ (X_1X_{2a})	$X_1 \times X_{2b}$ (X_1X_{2b})
4	9	1	0	9	0
5	11	0	1	0	11
6	10	0	0	0	0
7	4	1	0	4	0
8	9	0	1	0	9
9	4	1	0	4	0
10	8	0	1	0	8
11	10	0	1	0	10
12	12	0	1	0	12
13	5	1	0	5	0
14	3	0	0	0	0
15	12	0	0	0	0
16	9	0	0	0	0
17	10	0	1	0	10
18	8	0	1	0	8
19	1	1	0	1	0
20	8	0	0	0	0
21	11	0	1	0	11
22	4	1	0	4	0
23	6	0	0	0	0
24	10	1	0	10	0
25	1	0	0	0	0
26	4	0	1	0	4
27	3	1	0	3	0

The researcher then ran the regression with the independent variables years in labor force (X_1), the two dummy variables for academic degree (X_2), and the two new variables representing the interaction between years in labor force (X_1) and academic degree (X_2). More precisely, the X's in this analysis will be years in labor force (X_1), the "PhD" dummy (X_{2a}), the "JD" dummy" (X_{2b}), the interaction variable between years in labor force and the "PhD" dummy (X_1X_{2a}), and the interaction variable between years in labor force and the "JD" dummy (X_1X_{2b}). Regression results are shown below.

$R^2 = 0.803$
dffort = 21

	b	SE of b	t	p-value
Intercept (constant)	28.194	8.460	3.332	0.003
Years in labor force (X_1)	2.078	1.071	1.940	0.066
"PhD" dummy (X_{2a})	-19.756	11.249	-1.756	0.094
"JD" dummy (X_{2b})	-12.404	12.252	-1.012	0.323
Interaction between X_1 and X_{2a} (X_1X_{2a})	4.109	1.677	2.450	0.023
Interaction between X_1 and X_{2b} (X_1X_{2b})	3.592	1.475	2.435	0.024

There was a noticeable increase in R^2 from the previous regression, which informally suggests that the interaction may be valuable in improving the model's predictions. Formally, we can use the F-test to decide if the model with the interaction is a statistically significant improvement over the simpler model that had work experience (X_1) and the dummies representing academic degree (X_2). Sum of squared errors for the model including interaction terms (Model 1) is 2,238.194, and above it was noted that the model without interaction (which is now in the role of Model 0) had sum of squared errors 3,107.067. The interaction required a set of 2 new interaction terms, so that s (df for the numerator) is 2. df_1 (df for the denominator) is N – (# of X's in the full model with interactions) – 1 = 27 – 5 – 1 = 21. Then the calculation of F is

$$F = \frac{(SSE_0 - SSE_1)/s}{SSE_1/df_1} = \frac{(3,107.067 - 2,238.194)/2}{2,238.194/21}$$
$$= \frac{868.873/2}{2,238.194/21} = \frac{434.437}{106.581} = 4.076.$$

For df = (2, 21), this value of F gives $p < 0.05$ (or 0.032 from an online F-calculator or statistical software), so the interaction was statistically significant. The regression model with the interaction terms fit the sample data significantly better than the model without the interaction.

Because the interaction involves both of the X's (work experience, X_1, and academic degree, X_2), and the interaction is statistically significant, the researcher can sensibly interpret the b's even though the p-values associated with the underlying variables X_1, X_{2a}, and X_{2b} were not small. First consider the results in terms of the estimated effect of years in the labor force (X_1). The interaction means that there will be different effects of years in labor force for people with different academic degrees, and we can no longer simply report one number as *the* estimated effect of years in labor force on income.

With the interaction terms, the regression equation is

$$\hat{Y} = a + b_1X_1 + b_2X_{2a} + b_3X_{2b} + b_4X_1X_{2a} + b_5X_1X_{2b}.$$

To see the different effects of years in labor force depending on academic degree, it is helpful to rewrite, as before, the regression equation separately for each academic degree.

$$\begin{aligned}
\text{PhD}: \hat{Y} &= a + b_1X_1 + b_2(1) + b_3(0) + b_4X_1(1) + b_5X_1(0) \\
&= (a + b_2) + (b_1 + b_4)X_1. \\
\text{JD}: \hat{Y} &= a + b_1X_1 + b_2(0) + b_3(1) + b_4X_1(0) + b_5X_1(1) \\
&= (a + b_3) + (b_1 + b_5)X_1. \\
\text{BA}: \hat{Y} &= a + b_1X_1 + b_2(0) + b_3(0) + b_4X_1(0) + b_5X_1(0) \\
&= a + b_1X_1.
\end{aligned}$$

With the regression written separately for each degree, the estimated effect of years in labor force (X_1) on annual income (Y) can be summarized this way:

Degree	Estimated Effect of Work Experience
PhD	$b_1 + b_4$ (or, the b for years in labor force plus the b for the interaction between years in labor force and the PhD dummy) = 2.078 + 4.109 = 6.187.
JD	$b_1 + b_5$ (or, the b for years in labor force plus the b for the interaction between years in labor force and the JD dummy) = 2.078 + 3.592 = 5.670.
BA	b_1 (simply the b for years in labor force, because BA is the reference category) = 2.078.

The effect of years in labor force appears substantially greater for people with advanced degrees (JD or PhD) than it does for people who do not have an advanced degree (BA). In analyzing these sample data, an additional year in the labor force (X_1) is estimated to increase predicted income (Y) by \$5,670 for someone with a JD and by \$6,187 for someone with a PhD, compared with only \$2,078 for someone with a BA. There is greater "return" on additional experience for those with higher degrees. From the t-statistics for the interaction terms in the regression output, the differences in the effect of years in labor force between either PhD- or JD-holders and BA-holders are statistically significant. Note, however, that the difference in this effect between PhD- and JD-holders is not statistically significant. When the reference category is changed to PhD or JD, the t-test for the interaction term associated with this difference gives $p = 0.753$.

The other way to look at the interaction is to focus on how differences in predicted income ($\hat{Y}$) between people with different degrees change depending on the level of work experience being considered. For this, we can rewrite the equation to collect terms as coefficients for the dummy variables:

$$\hat{Y} = a + b_1X_1 + b_2X_{2a} + b_3X_{2b} + b_4X_1X_{2a} + b_5X_1X_{2b}$$
$$= a + b_1X_1 + (b_2 + b_4X_1)X_{2a} + (b_3 + b_5X_1)X_{2b}.$$

Using the b's from the table, the estimated differences between the academic degrees are as follows:

Comparison	Difference in Predicted Income
PhD vs. BA	$b_2 + b_4 X_1 = -19.756 + 4.109\ (X_1)$.
JD vs. BA	$b_3 + b_5 X_1 = -12.404 + 3.592\ (X_1)$.
PhD vs. JD	$(b_2 + b_4 X_1) - (b_3 + b_5 X_1)$ $= (b_2 - b_3) + (b_4 - b_5)\ X_1$ $=(-19.756 - -12.404) + (4.109 - 3.592)\ (X_1)$ $= -7.352 + 0.517\ (X_1)$.

For example, suppose we are interested in comparing the differences in predicted income among the academic degrees for people at 3 years ($X_1 = 3$) and at 5 years ($X_1 = 5$) of work experience. These estimated differences in predicted income among the academic degrees are obtained by using 3 and 5 as the value of X_1 in the above expressions.

At 3 years of work experience ($X_1 = 3$):

Comparison	Difference in Predicted Income
PhD vs. BA	$b_2 + b_4 X_1 = -19.756 + 4.109\ (X_1) = -19.756 + 4.109\ (3) = -7.429$.
JD vs. BA	$b_3 + b_5 X_1 = -12.404 + 3.592\ (X_1) = -12.404 + 3.592\ (3) = -1.628$.
PhD vs. JD	$(b_2 + b_4 X_1) - (b_3 + b_5 X_1) = (b_2 - b_3) + (b_4 - b_5)X_1$ $= -7.352 + 0.517\ (X_1) = -7.352 + 0.517\ (3) = -5.801$.

At 5 years of work experience ($X_1 = 5$):

Comparison	Difference in Predicted Income
PhD vs. BA	$b_2 + b_4 X_1 = -19.756 + 4.109\ (X_1) = -19.756 + 4.109\ (5) = 0.789$.
JD vs. BA	$b_3 + b_5 X_1 = -12.404 + 3.592\ (X_1) = -12.404 + 3.592\ (5) = 5.556$.
PhD vs. JD	$(b_2 + b_4 X_1) - (b_3 + b_5 X_1) = (b_2 - b_3) + (b_4 - b_5)X_1$ $= -7.352 + 0.517\ (X_1) = -7.352 + 0.517\ (5) = -4.767$.

From the calculation above, we can see that the differences in predicted income among academic degrees differ in important ways depending on what level of work experience we are considering. When work experience is 3 years, the difference in predicted income between PhD- and BA-holders is –7.429. That is, at 3 years in the labor force, a PhD is predicted to earn $7,429 *less* than a BA. However, this PhD-BA difference changes as work experience increases. When work experience is 5 years, the difference in predicted income between PhD- and BA-holders becomes 0.789, meaning that a PhD is predicted to earn $789 *more* than a BA at 5 years of work experience.

For JD vs. BA, the difference is –1.628, or $1,628 less for the JD than for the BA, at 3 years of work experience. However, when work experience is 5 years, the difference is 5.556, or $5,556 more for the JD than for the BA. For PhD vs. JD, the difference at 3 years of work experience is –5.801, or $5,801 less for the PhD than for the JD (this difference could also be obtained by calculating [–7.429 –1.628]). When work experience is 5 years, the calculation gives –4.767, or $4,767 less for the PhD than for the JD. (This difference could also be obtained from 0.789 – 5.556.)

Clearly all of these figures will change depending on the value of years in labor force (X_1) that is plugged in, so it is now rather complicated to describe the differences in predicted income between the different degrees. As one way to help make sense of the results, notice that in the calculations for comparing the academic degrees above, it happens that the first number is negative and the second (the coefficient on X_1) is positive; for example, the estimated PhD-BA difference was calculated as $-19.756 + 4.109\ (X_1)$. Of course in other analyses that might not turn out to be the case, but it is for these data. As X_1 is years of work experience, and so cannot have a negative value, it is straightforward to see that for some values of X_1 the result would be negative (here, indicating that BAs earn more), and for other values of X_1 it would be positive (indicating that PhDs earn more). In such a case, we can find the specific value of X_1 at which this result changes sign, which may be very helpful for interpretation.

We can find this value by setting the expression equal to 0 and then solving for X_1. For the PhD vs. BA comparison, we solve $-19.756 + 4.109\ (X_1) = 0$, which gives $X_1 = 4.808$ as the solution. That indicates that we estimate that for people with work experience (X_1) under 4.808 (or roughly 5) years, the predicted income is higher for a BA- than a PhD-holder, because for these values of X_1, $-19.756 + 4.109\ (X_1)$ will be negative. But above that level of work experience, the predicted income is higher for the PhD, and the gap will continue to grow as more years in the labor force are accumulated, because $-19.756 + 4.109\ (X_1)$ will be positive. For the JD vs. BA comparison, we similarly solve $-12.404 + 3.592\ (X_1) = 0$, giving $X_1 = 3.453$ as the solution. That estimate indicates that for people with work experience (X_1) under 3.453 (or roughly three and a half) years, predicted income is higher for a BA- than a JD-holder. But above that level of work experience, the predicted income is higher for the JD.

As these were hypothetical data, we do not need to think too much about these results, but the findings suggest a pattern in this hypothetical world in which the positions through which BA-holders enter the field of law tend to have higher starting salaries but less potential for salary growth than do PhD- and JD-holders. As work experience grows, the PhD- and JD-holders will on average increasingly outrun the BA-holders in salary.

To complete these comparisons for all pairs, the PhD vs. JD comparison, using $-7.352 + 0.517\ (X_1)$, shows that the value of X_1 at which the difference changes sign is $X_1 = 14.221$. Notice that this value is greater than any person's value of X_1 in the sample. It is important to note, then, that for the range of

X_1 values actually seen in the sample, the PhD vs. JD difference in predicted income is always negative, meaning that predicted income is higher for the JD at any observed level of work experience.

These expressions in the example all had the form *negative* + *positive* (X_1), but the same kind of ideas apply to the situation of *positive* + *negative* (X_1), except then the difference is positive for X_1 below the special value and negative for X_1 above it. Suppose the two terms are the same sign, as *negative* + *negative* (X_1) or *positive* + *positive* (X_1). If X_1 is, by its nature, restricted to positive values (which will be the situation for many variables that we use as independent variables, like years of work experience here), then the difference will always be negative (in the first case) or positive (in the second), regardless of the value of X_1.

4.4 Interaction Between Two Numerical Independent Variables

An interaction between two numerical independent variables follows the same logic that we explored above for interaction between a categorical and numerical variable: the effect of one independent variable on $\hat{Y}$ changes depending on the value of another independent variable. This leads to the same method of expressing an interaction in a regression equation, via a new term created by multiplying the two independent variables involved in the interaction. Interpretation is also quite similar.

Suppose that we are interested in investigating an interaction between numerical variables; for this discussion, say those are X_2 and X_3. As in the earlier presentation of interaction, we create a new variable by multiplying these two independent variables and include that in the regression along with the original X_2 and X_3. The regression equation for this example includes only one additional independent variable X_1, but of course there could be more. Then our regression equation is

$$\hat{Y} = a + b_1X_1 + b_2X_2 + b_3X_3 + b_4X_2X_3.$$

Statistical significance of the interaction is assessed by statistical significance of this additional term X_2X_3, using the p-value from the t-statistic on the output. Just as with the categorical-numerical interaction situation earlier, we can examine how the predicted Y changes as a result of changes in X_2 or X_3. Now, however, there is no distinction between the natures of the two variables, as both are numerical. Focusing on X_2 first, we can rewrite the regression equation to group the parts involving X_2:

$$\begin{aligned}\hat{Y} &= a + b_1X_1 + b_2X_2 + b_4X_2X_3 + b_3X_3 \\ &= a + b_1X_1 + (b_2 + b_4X_3)X_2 + b_3X_3.\end{aligned}$$

Viewed this way, we see that the estimated effect of X_2 on Y is $(b_2 + b_4 X_3)$. As expected from the logic of interaction, this effect of X_2 changes depending on what value we plug in for X_3. The same type of regression equation rewriting would apply when we focus on X_3, so the estimated effect of X_3 on Y turns out to be $(b_3 + b_4 X_2)$. In each case, b_4 indicates how the value of the other variable in the pair changes the effect of the variable on which we are focusing.

To look at this more concretely, suppose that X_2 represents years of education, and X_3 represents age, and that our estimated regression equation is

$$\hat{Y} = 19.7 + 3.3X_1 + 5.2X_2 - 2.9X_3 + 0.1X_2X_3.$$

Also suppose that the p-value from the t-statistic for the additional term $X_2 X_3$ did show statistical significance, so we believe that there is statistically significant interaction between X_2 and X_3 and will interpret the estimates here. As before, we will still do this even if we did not find statistical significance for the separate terms X_2 or X_3 in this regression.

Focusing on X_2 (years of education), the estimated effect of X_2 is $(b_2 + b_4 X_3) = 5.2 + 0.1\ X_3$. To make this more specific, we need to plug in a value of X_3. With X_3 representing age, perhaps we are especially interested in the effect of education on $\hat{Y}$ for a 50-year-old person. At that age, the effect of education is $(5.2 + 0.1 \times 50) = 5.2 + 5 = 10.2$, indicating the predicted Y for a 50-year-old increases by 10.2 units for each additional year of education, holding X_1 constant, with age implicitly held constant too. For a 30-year old, on the other hand, the effect of education would be smaller: $(5.2 + 0.1 \times 30) = 5.2 + 3 = 8.2$. The positive b_4 means that this estimated effect of education becomes greater—more positive—as age increases.

Now turning to X_3 (age), its estimated effect on $\hat{Y}$ is $(b_3 + b_4 X_2) = -2.9 + 0.1\ X_2$. Again, to go further we need to consider a specific value of X_2. Suppose we are interested in understanding the age effect for high school graduates and so set $X_2 = 12$. Then $(-2.9 + 0.1 \times 12) = -1.7$. For a high school graduate, the predicted Y decreases by 1.7 units for each additional year of age, holding X_1 (and of course X_2 also) constant. For a college graduate, this estimate is $(-2.9 + 0.1 \times 16) = -1.3$, so the relationship between age and Y is estimated to become less negative as education increases. Because b_3 and b_4 have opposite signs in this example, it seems that if education were large enough, the effect of age on the predicted Y would become positive. Note, however, that with these estimates, a change in the sign would require a value of more than 29 for education, and most surveys would not permit a respondent to enter such a large number for years of education.

Some analysts prefer *centering* the pair of numerical variables that will be used in an interaction. Centering refers to the act of subtracting a variable's mean from each of its values, so that its values become deviations from its mean rather than raw measures. Centering can be done in statistical software rather than by hand; we simply create a new variable calculated as $(X - \bar{X})$. With no interactions, none of the multiple regression results that we have discussed change when an independent variable (or variables) is centered. For the most part this remains true when centered variables are involved in an interaction, but certain regression output will change in that situation.

The estimate b for the interaction term (b_4 in our example above) will be the same whether the two variables in the interaction were centered or not. But in the terms for the separate variables, the estimates (b_2 and b_3 in the example) will change. In the example above, the estimated effect of X_2 is ($b_2 + b_4 X_3$), and we see that this will simply be b_2 if the value of X_3 is set at 0. So we could interpret b_2 as the change in $\hat{Y}$ resulting from a 1-year increase in education when age is 0. Of course a 0 value for age would surely be impossible in survey data about education; in this and many other instances, this interpretation of b_2 will not be helpful, because it is based on an impossible value for X_3. But if we are working with centered versions of these variables, then a 0 value for X_3 is legitimate, because, in the deviation form implied by centering, a 0 value refers to the mean age in the sample. In that case we can interpret b_2 as the change in $\hat{Y}$ when increasing years of education by 1 for a person at the sample mean of age. So in this example, centering would allow for a more natural interpretation of b_2 on its own than would have been possible without centering.

The other motivation for centering is to avoid collinearity. (This motivation will be discussed again in a different setting in Chapter 6.) For uncentered variables, especially if they take only positive or only negative values, X_2 and X_3 could be rather highly correlated with the new term $X_2 X_3$. Centering is a way to avoid this possibly high correlation, with the multiplication now involving both negative and positive numbers.

In comparing regression results using the centered variables and their interaction to those using the original variables, one sees that many parts of the results are unchanged, including R^2, b's, standard errors, t-statistics, p-values for other variables, and the b and standard error for the interaction term $X_2 X_3$. However, in the presence of the interaction term, b's and standard errors for X_2 and X_3 will be different between the analyses with and without centering. These model estimates differ in a known way between the uncentered and centered analyses, so this apparent difference is cosmetic rather than fundamental. But because those b's are typically interpreted along with b_4 (corresponding to the interaction term $X_2 X_3$) even if they show large p-values, this response to possible collinearity may not be so important in practice. Centering is therefore not a universal strategy when working with interactions between numerical independent variables.

4.5 Interaction Between Two Categorical Independent Variables

Finally, we look at interaction between two categorical independent variables. As with the other interaction situations, we will create new variables through multiplication. The number of these new variables could be quite large, however, if the two variables have many categories. That is because we will create the new variables by multiplying all pairs of dummy variables, meaning one in the pair from the set created for the first categorical independent variables and the other from the set created for the second categorical variable, and doing this for all such possible pairs. Although we will discuss this approach in terms of multiple regression, this sort of interaction is also closely related to the ANOVA framework noted earlier in Section 3.8.3.

When a categorical independent variable is dichotomous (has just two categories), it can be represented by a single dummy variable. If two dichotomous categorical variables are involved in an interaction, there is just a single pair of dummies, so only one new variable needs to be created to represent the interaction. Suppose our data include variables showing political party (Republican or Democratic) and employment status (employed or unemployed), with X_2 being a dummy variable in which Republican = 1 and X_3 being a dummy in which employed = 1. Then the interaction between political party and employment status will be represented by a single new variable $X_2 X_3$. Because X_2 and X_3 are both dummies, taking on only the values 0 or 1, the new $X_2 X_3$ will also always be 0 or 1. When both X_2 and X_3 are 1 (for a Republican who is employed), then $X_2 X_3$ will be 1. But for any other combination of party and employment status, $X_2 X_3$ will be 0, as at least one of the two values being multiplied will be 0. The interaction requires only a single new term, so the usual t-statistic and p-value for that variable can be used to assess statistical significance of the interaction.

If our regression equation is $\hat{Y} = a + b_1 X_1 + b_2 X_2 + b_3 X_3 + b_4 X_2 X_3$, we can understand the meaning of the interaction by examining the part of the equation involving X_2 and X_3, ignoring X_1 for the moment. The possible combinations of political party and employment status can be represented in a simple 2 × 2 table, with the cell values indicating the relative value of $\hat{Y}$ for a person with that combination of the two variables. Again this is holding X_1 constant, and to get the actual rather than relative prediction we would also need to add the part of the regression equation for $(a + b_1 X_1)$. We obtain these relative values of $\hat{Y}$ by setting the dummy variables in the regression equation segment $b_2 X_2 + b_3 X_3 + b_4 X_2 X_3$ to 1 or 0 as appropriate for the particular cell. For example, the part of $\hat{Y}$ given by $b_2 X_2 + b_3 X_3 + b_4 X_2 X_3$ will equal $b_2 (1) + b_3 (1) + b_4 (1 \times 1)$, or $b_2 + b_3 + b_4$ for an employed Republican, $b_2 (0) + b_3 (1) + b_4 (0 \times 1) = b_3$ for an employed Democrat, and so on.

		Employment Status	
		Employed	**Unemployed**
Political Party	Republican	$b_2 + b_3 + b_4$	b_2
	Democratic	b_3	0

Looking down the columns, we see the impact of the interaction, as the difference in predicted Y between Republicans and Democrats depends on employment status. For employed people, the political party difference, holding X_1 constant, is $b_2 + b_4$, from $(b_2 + b_3 + b_4) - b_3$. For unemployed people it is simply b_2, from $b_2 - 0$. Looking across the rows, the difference in predicted Y between employed and unemployed people also depends on political party. Among Republicans, this difference is $b_3 + b_4$, from $(b_2 + b_3 + b_4) - b_2$, but among Democrats, it is b_3. The estimate b_4, associated with the interaction term, is the key to the comparison of differences in each case.

Suppose that our regression output gave an estimated regression equation of

$$\hat{Y} = 14.7 - 0.8X_1 + 1.3X_2 - 1.8X_3 + 0.6X_2X_3,$$

with the p-value for the interaction term indicating statistical significance. Then these b's can be used to show the estimated differences among categories of political party and employment status, holding X_1 constant, as in the above table. Because of the interaction, we must to be careful to specify which category of the other variable involved in the interaction is being used as the basis for the interpretation. Focusing first on X_2 (political party), the difference in the predicted Y between Republicans and Democrats is $(b_2 + b_4) = (1.3 + 0.6) = 1.9$ units for employed people, and $b_2 = 1.3$ units for unemployed people, holding X_1 constant. In either employment status, the predicted Y is greater for Republicans than Democrats, but this difference by party is greater among the employed. Turning to employment status, the difference in $\hat{Y}$ between employed and unemployed people is $(b_3 + b_4) = (-1.8 + 0.6) = -1.2$ for Republicans, and $b_3 = -1.8$ for Democrats, holding X_1 constant. In either party, the predicted Y is less for employed people than for unemployed people, but the difference by employment status is greater among Democrats than among Republicans.

When the categorical variables have more than two categories, and therefore yield more than one dummy variable, the same construction of interaction terms is made, except that now there may be many more of these new terms needed. For example, suppose our survey data had five categories for race/ethnicity (X_2) and six categories for educational attainment (X_3). With one less dummy than the number of categories, race/ethnicity will be represented by the set of four dummies X_{2a}, X_{2b}, X_{2c}, and X_{2d}, and educational attainment will be represented by five dummies X_{3a}, X_{3b}, X_{3c}, X_{3d}, and X_{3e}. Each of the possible combinations of a dummy from one set and a dummy from another set will be turned into interaction terms: $X_{2a}X_{3a}$, $X_{2a}X_{3b}$,..., $X_{2a}X_{3e}$, $X_{2b}X_{3a}$,..., $X_{2b}X_{3e}$, ..., $X_{2d}X_{3a}$,..., $X_{2d}X_{3e}$. In all there will be 4 × 5 = 20 such terms. An F-test comparing the regression model with and without this set of 20 terms will be used for an overall assessment of statistical significance of the interaction, while the t-statistics and p-values for these terms on the output can be used to test statistical significance of specific differences between differences. Obviously great care will need to be taken in interpreting all of the possible category differences in such a case. But even though there are many more of these interaction terms, a table like that for the example above still can be created, except that its size will be 5 rows × 6 columns (for the five categories of race/ethnicity and the six categories of educational attainment).

4.6 Exercises

4.6.1 Exercise 1: Sentence Length for Convicted Robbery Offenders Continued

Using the data from Section 3.7.1 (p. 90, Chapter 3), conduct a multiple regression analysis that can help test the claim that age's effect on severity of punishment differs among socioeconomic classes (with prior criminal record controlled). Fully interpret the results.

4.6.2 Exercise 2: Public Transportation and Health Benefit

The availability of public transportation and its health benefits are often discussed. At the city level, Dr. Lyons hypothesizes that the effect of public transportation accessibility (measured by a public transit score, varying between 1 and 100) on health (measured as the population obesity percentage) changes depending on the city's weather. Weather is measured with three categories, reflecting three levels of winter conditions (mild, moderate, and severe). He used "severe" as reference category, so that X_1 = public transit score, X_{2a} = "mild" dummy, and X_{2b} = "moderate" dummy.

1. Write a regression equation (for the dependent variable obesity percentage) that includes an interaction between public transit score and weather.
2. What is the estimated effect of public transit score for a city with mild winter weather? Express it using the equation you wrote in Question (1).

4.6.3 Exercise 3: Interaction Between Two Numerical Variables

A researcher is analyzing data that include a happiness scale and is using happiness as the dependent variable in a multiple regression. Among the independent variables are age and years of education and an interaction between them. (Both variables are uncentered.)

Suppose that the interaction term was statistically significant and that the b's for age, education, and the interaction term were –3.32, 7.85, and 0.16, respectively. Use this information to (a) express the estimated effect of age on happiness as a function of education, (b) express the estimated effect of education on happiness as a function of age, and (c) find the estimated effect of age on happiness for a college graduate (16 years of education).

4.6.4 Exercise 4: Interaction Between Two Categorical Variables

In a research project on fear of crime, type of residence, categorized as (urban, suburban, and rural), and political party identification, categorized as (Democratic, Republican, any minor party, and independent), are going to be used as independent variables. The researcher's theory also suggests that there should be an interaction between type of residence and party identification.

1. Write out the regression equation involving these two variables and their interaction.
2. Use a chart like those shown in this chapter to report the expression for the relative predicted value of Y for each combination of type of residence and party identification.

5

Logged Variables in Multiple Regression

We studied the basic format of multiple regression in Chapter 2 and extended that with categorical independent variables in Chapter 3 and interaction terms in Chapter 4. In this chapter and Chapter 6, we examine some additional key extensions, involving particular transformations of numerical variables that give multiple regression greater flexibility in capturing the nature of relationships between independent and dependent variables. We begin by considering the *logarithmic* (usually shortened to *log*) transformation, including the motivation for that transformation, some core properties of logarithms, and use of the transformation in multiple regression. We then discuss the interpretation of multiple regression results under three possible scenarios: (a) Y is logged, but X is not, (b) X is logged, but Y is not, and (c) both X and Y are logged, followed by a full example. The chapter concludes with exercises.

5.1 Why Log a Variable?

The most common motivation for log-transforming (or, more simply, logging) a numerical variable before using it in a multiple regression is that we believe this transformation better captures the inherent nature of the concept that the variable aims to measure. By logging a variable we are saying something fundamental about the variable's meaning and the information that it conveys. In particular, we are asserting that for this variable it is more meaningful to focus on *multiplicative*, rather than *additive*, differences in its values.

A brief example helps explain this. Suppose that we are studying adults' annual incomes, and we are comparing incomes in two pairs of people, with initials (A, B) and (C, D). A and B work at jobs with rather modest incomes, while C and D are a professional athlete and a movie star. Further suppose that these four individuals' annual incomes are as follows:

A: $20,000

B: $30,000

C: $10,000,000

D: $10,010,000

Now compare the incomes within each pair. If we conceive of differences between incomes in terms of addition and subtraction—that is, in additive terms—simple subtraction shows us that there is the same difference of $10,000 between the higher and lower earner in each pair. But when we consider the nature of income, and in particular its relation to lifestyle, this seems misleading. In the context of the relatively modest annual salaries earned by A and B, $10,000 is likely a pretty important difference. B can probably afford a noticeably better residence or vehicle than A and in general enjoy a visibly more comfortable lifestyle. But for the very high earners C and D, it is likely impossible to tell from their lifestyles who earns more. In that context of exceptionally well-off people, $10,000 seems like a very small difference, almost just a rounding error when viewed against the pair's $10 million incomes.

In this example, then, the usual additive thinking does not seem to correctly capture the nature of the comparison. However, shifting to a multiplicative view of the differences appears to help. In the first pair, the higher earner's income is (30,000/20,000) = 1.50 times greater than the lower earner's. Expressed slightly differently, B's income is 50% greater than A's. The language of percentages may feel more intuitive than that of raw multiplicative differences, and we can convert a multiplicative difference to a percentage by subtracting 1 and then multiplying by 100: (1.50 – 1) × 100 = 50%.

But in the second pair, the higher earner's income is (10,010,000/10,000,000) = 1.001 times greater than the lower earner's. In percentage terms, D's income is 0.1% (one-tenth of 1%) greater than C's. Note that these ratios B/A or D/C would be exactly 1 if the two incomes in the pair were the same, so in this multiplicative view "1 times greater" corresponds to no difference. 1.50 times (or 50% higher) is therefore a much bigger multiplicative difference than 1.001 times (or 0.1% higher). In this view, the disparity between A and B is of a much greater magnitude than the disparity between C and D, which seems to be a better fit with our intuition about this situation. In this case, the multiplicative framework seems very appealing.

A slightly different way to express our intuition is to note that for income, the same additive difference (here $10,000) has more real meaning when it occurs at relatively low incomes (as in pair A–B) than when it occurs at high incomes (like pair C–D). When the same additive difference is more meaningful at lower values of the variable, we are again led to take a multiplicative view of the differences. For income, then, this multiplicative perspective seems more satisfactory and realistic than an additive view.

Logging income is a way to incorporate this multiplicative view into our analysis. That is because additive thinking about a logged variable is equivalent to multiplicative thinking about the original, untransformed variable; if we use the logged version of income in a multiple regression, we bring a multiplicative interpretation to the results for that variable. This desire to take a multiplicative view of differences in a variable's values will be a main motivation for logging. Of course in many cases, an additive view of a variable's differences will seem perfectly appropriate, and a given additive difference may seem to have the same meaning over the whole range of the variable. In such cases we will not need to log, and it is important to be clear that many studies do not include any logged variables. But the availability of the log transformation brings a good deal of flexibility to multiple regression, enhancing our ability to more accurately represent underlying social science concepts.

This discussion has focused on substantive arguments for taking a multiplicative view of a variable's differences via the log transformation. In some cases, though, variables in multiple regression are logged on the basis of inspection of the data and regression results, rather than through this sort of more theoretical thinking. After running a regression, analysts may inspect various plots involving the errors/residuals ($Y - \hat{Y}$) for indications that variables in the regression should be logged. For instance, logging may be helpful if the errors' distribution does not appear much like the expected bell-shape of the normal distribution. Or in a plot of the residuals against the observed dependent variable Y or the predicted values $\hat{Y}$, a pattern of errors seeming to be related to the values of the dependent variable, or of larger absolute values being associated with larger values of the dependent variable, can sometimes be addressed by logging. The analyst can experiment with logging variables, comparing such plots before and after to see what impact the transformation had.

We do not discuss such residual plots further here, but statistical software generally will offer a menu of these plots, and more advanced texts can provide more detail on their interpretation. Simple plots of an X against Y may also seem to show a relationship that suggests logging one or both of the variables. (Chapter 6 includes figures illustrating various X–Y relationships involving logs.) Here we instead emphasize how our understanding of a variable's nature (whether or not the same additive difference is more meaningful at lower values of the variable) can help guide the decision to log or not. Still, the choice to log a variable for use in multiple regression will sometimes be made on practical rather than theoretical grounds. In any case, appropriate logging can help many aspects of the regression analysis work "better" and yield more insightful interpretations.

5.2 Main Properties of Logarithms

Before discussing issues around the use of logged variables in multiple regression, we need to establish some of the key properties of logarithms. Our goal is simply to provide a solid foundation for understanding regression using logs, so we highlight a limited number of facts about logs that are most helpful in the regression context. Of course there are many more ideas surrounding logs that are important for other purposes, but the full mathematical richness of logarithms is not needed in order to correctly interpret regression results involving logged variables.

First, a definition: after specifying which *base* is being used, the logarithm of a number is simply the exponent, or power, that we need to apply to the base in order to return that number. For example, one common choice is *base 10*. With base 10, the log of 100 is 2, because $10^2 = 100$, and the log of 1,000 is 3, because $10^3 = 1{,}000$. The general definition is that for a base c and a number d, $c^{\log d} = d$: that is, the log of d is the number that, when used as a power of c, returns d. Note that the log will not necessarily be a whole number; for example, 225 is between 100 and 1,000, so its base 10 log is between 2 and 3. From a calculator, the base 10 log of 225 is (rounded) 2.352, with $10^{2.352} = 225$. Because $c^{\log d} = d$, the operation in which c is raised to the power log d is sometimes called the *antilog*, as it "undoes" the log operation and returns the number d ($10^{2.352} = 225$ in the example here).

Base 10 is intuitive, but in statistical applications it is more common to use the base e, or the *natural log* (usually shown as "ln" on a calculator). e, sometimes called Euler's number, is one of the specific numbers that have a prominent role in mathematics, like π is a special number in many areas of math, and e is defined for its first 12 decimal places as e = 2.718281828459. (As with π, the value of e has been determined to millions of decimal places.) One reason why e is often used as the base for logarithms in statistical applications is that when a number d is only slightly above or below zero, the base e logarithm of (1 + d) will be very close to d, and that can be helpful for certain interpretations. We discuss that further in the appendix (see Section 5.7). Although we do not explore other reasons for this choice, we will use the natural log (base e) throughout this chapter. Note that the core ideas about logs and the ways that they allow us to represent multiplicative thinking about a variable's differences do not change if some other base is used, so it is straightforward to modify our discussion for some other base if desired. The form of the antilog (undoing the log) is still the same with base e. For base e, we have $e^{\log d} = d$. If d is 2, then $e^{\log 2} = e^{0.693} = 2$.

We now list some important properties of logs and exponents and illustrate them with brief examples.

1. *The logarithm is not defined for 0 or a negative number.* Note that the result of a log operation can be zero or negative. For instance, log 1 = 0 (because $e^0 = 1$; by definition any number raised to the power of 0 equals 1), and log 0.5 = −0.693 (again we are using the natural log, which may be shown as "ln" on a calculator: $e^{-0.693} = 0.5$). But the number that we are trying to log cannot be negative or zero; a calculator will indicate an error in that case. So all of our discussion assumes that any number being logged is > 0.

   There may be instances in which we believe that a variable "should" be logged (that is, from a theoretical perspective its differences should be considered in multiplicative rather than additive terms), but we notice that our sample includes zero or negative values for it. If so, we cannot log it, even if we otherwise think there is a good argument for logging. One alternative would be to slightly adjust our data to remove zero or negative values. For instance, adding a small amount to each person's value of a variable might eliminate any zero values of that variable. In the case of income, this could mean treating anyone with no income as if their income were instead $1. This seems like a very minor distortion of the actual information and would allow income to be logged in the regression analysis.

This approach can be sensible sometimes, but in general we should be cautious about making any adjustments in our raw data. We would need to be very clear in our report or paper that we had made this adjustment, and ideally we should experiment with adjustments of different size, in order to be sure that our final regression results are not being unduly influenced by the precise adjustment that we used. Although this sort of adjustment can be a practical tool in some cases, the main idea to remember is that variables with zero or negative values cannot be logged.

2. *The log of a product is the sum of the logs.* This means that if we want the log of the product (that is, the result of a multiplication) of two numbers, we can find each number's log and add those. Writing this in equation form, where f and g are positive numbers, this means that $\log (f \times g) = \log f + \log g$. For example, the log of 5×6 will equal log 5 plus log 6 [$\log (5 \times 6) = \log 5 + \log 6$]. We can confirm this with a calculator: $\log 5 = 1.609$, and $\log 6 = 1.792$ (both rounded), which sum to 3.401. To check that the relationship holds, note that $5 \times 6 = 30$, and use a calculator to confirm that log 30 does indeed equal 3.401. (Rounding could have produced a small discrepancy, but it happened not to in this instance.) The same sort of property holds for division, but with subtraction: $\log (f/g) = \log f - \log g$.

   One result of this property is that an additive increase of one in the log is equivalent to a multiplicative increase of e (assuming that we are using the natural log) in the original number. Using the rule above, $\log (g \times e) = \log g + \log e$, and it is also true that $\log e = 1$ (because $e^1 = e$; any number to the power one is just that number). So $\log (g \times e) = \log g + 1$, which confirms that multiplying g by e adds one to its log. Here we presented this for the natural log and base e, but whatever base we use, adding one in the log scale is equivalent to multiplying the underlying number by the base.

3. *The log of a number raised to an exponent is that exponent times the log of the number.* This means that $\log (f^g) = g \log f$. When the exponent is a whole number, we can see how this follows from the property above: for example, if $g = 3$, $\log (f^3) = \log (f\ f\ f) = \log f + \log f + \log f = 3 \log f$. But the property also generalizes beyond whole numbers. For example, the rule says that $\log (3.75^{1.83}) = 1.83 \log 3.75$, and we can confirm this with a calculator. First, $\log 3.75 = 1.322$, and $1.83 \times 1.322 = 2.419$. Second, $3.75^{1.83} = 11.233$, and $\log (11.233) = 2.419$, so the relationship holds. (Again there could have been a small discrepancy due to rounding.)

Because exponents (or powers) are so closely related to logs, it is helpful to also include two of the general rules governing exponents. These will be helpful in explaining some of our regression interpretations for logs.

4. *The product of powers of the same base is the base to the sum of the powers.* This can be expressed simply as $f^g \times f^h = f^{g+h}$. For example, $3^2 \times 3^4 = 3^6$ from the rule; $3^2 = 9$ and $3^4 = 81$, and $3^2 \times 3^4 = 9 \times 81 = 729$. From the calculator, $3^6 = 729$, which agrees with $3^2 \times 3^4$.

5. *For the power of a power, multiply the exponents.* This rule says that $(f^g)^h = f^{(g \times h)}$. As an illustration of this rule, consider $(2.5^4)^3$. From the rule, $(2.5^4)^3 = 2.5^{4\times3}$. The calculator shows first that $2.5^4 = 39.0625$, so $(2.5^4)^3 = 39.0625^3 = 59{,}604.645$ (rounded to three decimal places). On the calculator we can also confirm that $2.5^{4\times3} = 2.5^{12} = 59{,}604.645$.

5.3 Using Logged Variables in Multiple Regression

With these fundamental properties of logs and exponents in hand, we can understand interpretations of regression results when logged variables are being used. First, note that if we do decide to log a variable, it will be straightforward to do so in our statistical software. We will be able to create a new variable (and give it an appropriate name) for which each case's value is the log of the value in the original variable. We are fortunately not required to laboriously carry out all that logging one case at a time with our calculator. As we will see below, however, we will still do some of our own calculations when interpreting regression results that involve logged variables.

Our discussion above examined why we would be inclined to log a variable when a multiplicative perspective on changes in that variable seems more meaningful than an additive perspective. This idea carries over to interpretation, and we will end up making a multiplicative interpretation of change in any variable in our regression that has been logged. Even so, many elements of interpretation that we learned in previous chapters do not change when logged variables are involved. In particular, R^2 still refers to the quality of our predictions of the regression's dependent variable, or the proportion of variation in the dependent variable that is explained by the regression's independent variables. Assessment of statistical significance, via t-statistics and p-values, also works the same way even if logged variables are involved.

Interpretations of b do change for logged variables, so we focus on those changes in this section. We will look at the interpretation of b under three scenarios: (a) Y is logged, but X is not, (b) X is logged, but Y is not, and (c) both X and Y are logged. For now we will assume that the t-tests, or F-tests for sets of dummies, showed statistical significance for all of the independent variables being discussed, so that we have moved ahead to interpreting b's. Of course in our actual analyses we will still follow the procedure of assessing statistical significance before interpreting any b.

5.3.1 Y is Logged, X is Not

We first consider the scenario in which Y has been logged, and we are interpreting b for an X that has not been logged. The regression equation with logged Y and two X's (of course there could be additional independent variables) is as follows:

$$\widehat{\log Y} = a + b_1X_1 + b_2X_2.$$

Notice that we wrote the left-hand side as "(log Y)-hat," rather than "log (Y-hat)," in order to emphasize that the logged version of Y is being used in our software's calculation of regression results. That is, it is not as if we carried out a regression with the "plain" Y and then after the fact logged the predicted values. Rather the logged Y is an inherent part of the regression analysis. However, it is difficult to be completely precise or consistent with this sort of notation, and it will sometimes be more convenient not to be too strict with it.

Of course we could simply leave this regression equation as is and be clear that in our interpretation of b_1 we are reporting the change in the predicted log Y when X_1 increases by one unit, holding X_2 constant. Similarly, b_2 indicates the change in the predicted log Y when X_2 increases by one unit, holding X_1 constant. This would be technically correct, but it is clearly an unsatisfactory way of communicating our regression results to others. Most people, even experienced researchers, are not very used to, or comfortable with, thinking about logged values and thus will find it difficult to grasp the meaning of results that are reported in those terms. Our results will be much more understandable if we discuss independent variables' effects on Y rather than their effects on log Y. But because Y is logged, we must report changes in Y in multiplicative, rather than additive, terms; in the current scenario, changes in X (which has not been logged) will still be considered in additive terms.

The regression equation above was $\widehat{\log Y} = a + b_1X_1 + b_2X_2$. To bring our interpretations back to Y, rather than log Y, we need to "undo" the log from $\widehat{\log Y}$ by using the antilog, or $e^{\widehat{\log Y}} = \hat{Y}$. (The antilog also lets us determine $\hat{Y}$ for a given set of X values: we calculate the predicted log Y just as we had calculated the predicted Y before, and then as the final step use the antilog to convert $\widehat{\log Y}$ into $\hat{Y}$.) We need to perform this exponentiation on both sides of the regression equation, leaving us with $\hat{Y} = e^{(a+b_1X_1+b_2X_2)}$. Applying the idea shown in rule #4 above, the right-hand side can be rewritten as a series of terms multiplied together, so that the rewritten regression equation is $\hat{Y} = e^a e^{b_1X_1} e^{b_2X_2}$.

Now imagine that the value of X_1 increases—additively, as X_1 has not been logged—by one unit. Then the regression equation would change to

$$\begin{aligned}\hat{Y} &= e^a e^{b_1(X_1+1)} e^{b_2X_2} \\ &= e^a e^{b_1X_1+b_1} e^{b_2X_2}.\end{aligned}$$

Again invoking rule #4, we have $\hat{Y} = e^a e^{b_1X_1} e^{b_1} e^{b_2X_2}$. If we compare this to how the regression equation looked before increasing X_1 by one unit ($\hat{Y} = e^a e^{b_1X_1} e^{b_2X_2}$), we notice that after increasing X_1 by one unit the multiplication now includes an extra term e^{b_1}. This means that the effect of an increase of one unit in X_1, holding other X's constant, is to multiply $\hat{Y}$ by e^{b_1}. This is the multiplicative effect on $\hat{Y}$ of an additive increase in X_1; again our interpretation involves a multiplicative change in the logged variable (in this case Y) and an additive change in the unlogged variable (in this case X_1).

As before, we may want to consider something other than a one-unit increase in X_1 when interpreting the regression results. Suppose that X_1 increases by c units (a negative value of c would correspond to a decrease in X_1). If we work through the same steps as we just did for an increase of one unit, we will find that the multiplicative effect on $\hat{Y}$ of this increase of c units in X_1 is $e^{b_1 c}$. Alternatively, we can use the rule #5 for exponents to note that this is equivalent to $(e^{b_1})^c$.

Recalling the discussion in Chapter 2, a main reason for interpreting some change other than a one-unit increase in X_1 is that it will help in assessing real-world significance of X_1's effect. Note that Y's part in the assessment of real-world significance changes when it has been logged. Our assessment of the magnitude of the change in Y could still take advantage of the range of Y, but we need to think of this range in multiplicative terms. That is, perhaps the largest value of Y in our sample is three times greater than the smallest, which would offer some context for understanding the size of the multiplicative change in $\hat{Y}$ that our results show. Even without considering the multiplicative range of Y, the magnitude of a multiplicative change, especially when expressed in percentage terms, can be more readily assessed than is the case for an additive change. Statistics like the sample mean of Y will not be helpful in assessing this magnitude, as those likely do not really help us understand the size of a multiplicative change in $\hat{Y}$.

When X_1 is a dummy variable, it is of course not sensible to talk about a one-unit increase in it. Then the interpretation of e^{b_1} will be as a multiplicative difference between $\hat{Y}$ for the named dummy category and $\hat{Y}$ for the reference category, holding other X's constant. That is, with a dummy variable X_1, $\hat{Y}$ for the named category is e^{b_1} times $\hat{Y}$ for the reference category. Even when X is a dummy variable, the fact that we logged Y means that we must make a multiplicative interpretation of the difference in $\hat{Y}$ implied by b.

An example with numbers may be helpful. We will simply assume statistical significance throughout the example, but if this were a real analysis we would, as usual, check that prior to interpretation of b's. Suppose that Y is logged, X_1 and X_2 are numerical variables (not logged), and X_3 and X_4 make up a set of two dummy variables, representing a three-category independent variable. Suppose too that our output indicates that the estimated regression equation was

$$\widehat{\log Y} = 4.627 + 0.126X_1 - 0.104X_2 + 0.208X_3 - 0.337X_4.$$

Beginning with b_1, a one-unit increase in X_1 will, holding other independent variables constant, multiply $\hat{Y}$ by $e^{0.126}$, which our calculator shows is 1.134. Another way to express this result is that a one-unit increase in X_1 increases $\hat{Y}$ by 13.4%, obtained by (1.134 – 1) × 100. It makes sense that this positive b resulted in a multiplicative effect > 1, as a multiplicative change of more than 1 corresponds to an increase in $\hat{Y}$ (remember that because Y was logged, we can be sure that it does not involve any negative values).

If we were interested in an increase of five units in X_1, again holding other independent variables constant, the resulting multiplicative change in $\hat{Y}$ would be $e^{0.126 \times 5}$, or $e^{0.630} = 1.878$. Rule #5 says that this will be the same as $(e^{0.126})^5$, and the calculation $(e^{0.126})^5 = 1.134^5 = 1.875$ confirms this, ignoring a small discrepancy due to rounding when limiting the result of $e^{0.126}$ to three decimals. Our interpretation is that a five-unit increase in X_1, holding other independent variables constant, multiplies $\hat{Y}$ by 1.875. This corresponds to an 87.5% increase in $\hat{Y}$.

Notice that b_2 is negative: $b_2 = -0.104$. Then a one-unit increase in X_2 (holding other X's constant) multiplies $\hat{Y}$ by $e^{-0.104} = 0.901$. Another way to express this effect is as a 9.9% decrease in $\hat{Y}$, obtained from $(0.901 - 1) \times 100 = -9.9\%$. When e^b is less than 1, the multiplication implies a decrease in the value of $\hat{Y}$, consistent with the negative b_2. If we are interested in the effect of a 10-unit increase in X_2, we could calculate $e^{(-0.104 \times 10)} = e^{-1.04} = 0.353$. This says that a 10-unit increase in X_2 multiplies $\hat{Y}$ by 0.353, which is equivalent to a 64.7% decrease. We could also obtain this result from $(e^{-0.104})^{10} = 0.901^{10} = 0.353$, again drawing on rule #5.

Next we will interpret the results for the two dummy variables. $b_3 = 0.208$, with $e^{0.208} = 1.231$. As X_3 is a dummy variable, our interpretation examines the difference in $\hat{Y}$ between the named dummy category and the reference; Y is logged, so this is a multiplicative difference. Holding other independent variables constant, $\hat{Y}$ for the category named in the dummy X_3 is 1.231 times greater than for the reference category, or 23.1% greater. X_4 is also a dummy, and its b is negative (−0.337). $e^{-0.337} = 0.714$, so $\hat{Y}$ for the category named in X_4 is 0.714 times $\hat{Y}$ for the reference category; this can also be expressed as $\hat{Y}$ for the category named in X_4 being 28.6% lower than $\hat{Y}$ for the reference category.

Suppose that we are interested in predicting the value of Y for a case in which $X_1 = 6$, $X_2 = 4$, and $X_3 = 1$ (which also implies that $X_4 = 0$, as X_3 and X_4 are a set of two dummy variables representing a single categorical independent variable). It is easy to plug these values into the equation for the predicted log Y:

$$\begin{aligned}\widehat{\log Y} &= 4.627 + 0.126(6) - 0.104(4) + 0.208(1) - 0.337(0) \\ &= 5.175.\end{aligned}$$

To return this predicted log Y to the original scale of Y, rather than the log, we can apply the antilog to obtain $\hat{Y}$: $e^{5.175} = 176.80$. This seems like a much easier calculation than working through $\hat{Y} = e^a e^{b_1X_1} e^{b_2X_2} e^{b_3X_3} e^{b_4X_4} = e^{4.627} e^{0.126 \times 6} e^{-0.104 \times 4} e^{0.208 \times 1} e^{-0.337 \times 0}$, but either will arrive at the same final predicted value (ignoring any small difference due to rounding).

Although the process of finding the predicted value of Y is straightforward even in the presence of a logged dependent variable, we still must remember that we cannot try to use it to seemingly find a general additive effect of an X on Y. For example, we might think we could next plug in $X_1 = 7$ (one unit greater than the X_1 value that we used before), leaving other X values the same, and the resulting $\hat{Y}$ would let us see the additive change in $\hat{Y}$ when X_1 increased by one unit and other X's were held constant. This would be incorrect; the problem is that this additive change in $\hat{Y}$ after a one-unit increase in X_1 would be specific to this particular set of X values. We would obtain a different result if we started with some other X values, such as $X_1 = 8$, $X_2 = 2$, $X_3 = 0$, and $X_4 = 1$ and then increased X_1 by one unit. In order to make general statements in this log Y scenario about the effect of a change in X on Y, by which we mean statements that will be correct no matter what X values are involved, we must stick with the multiplicative interpretation discussed earlier.

We noted earlier that assessment of real-life significance is somewhat simpler when Y has been logged. We still need to be concerned with whether a one-unit increase in X provides a reasonable basis for making this assessment, but, as noted above, the resulting multiplicative change in $\hat{Y}$ can be evaluated without referring to Y's sample mean. We still may want to consider the observed range of Y, but in multiplicative terms; for instance, it may be that the largest value of Y in the sample is 4.2 times greater than the smallest value of Y. That can provide helpful context when interpreting the multiplicative change in $\hat{Y}$ corresponding to a one-unit increase in X. For the most part, however, we can interpret the magnitude of a multiplicative change in $\hat{Y}$, especially when expressed as a percentage change, without appealing to Y's descriptive statistics. A researcher could usually evaluate, for example, a 15% increase in $\hat{Y}$ without drawing on the descriptive statistics of Y.

5.3.2 X is Logged, Y is Not

In this scenario, the dependent variable Y has not been transformed, but one (or more) of the independent variables X has been logged. Because Y has not been logged, our interpretations will involve additive changes in Y, but interpretations for an X that has been logged will be made in terms of multiplicative changes in that X. Our regression equation (here with two X's) now looks like $\hat{Y} = a + b_1 X_1 + b_2 \log X_2$. X_1 is untransformed, so the usual interpretation of b_1 holds, while X_2 has been logged.

If we simply read the regression equation as it is, an increase of one in the value of log X_2 will (additively) change $\hat{Y}$ by b_2, holding other independent variables constant. This is technically correct, but an interpretation that refers to a change in log X_2 will not be very helpful in communicating our findings to others. We would rather that the interpretation be reported in terms of changes in X_2; we can do this, but because X_2 has been logged, our interpretation must involve multiplicative, not additive, changes in X_2.

To make this shift, remember that our rule #2 for logs implies that $\log (f \times e) = \log f + \log e = \log f + 1$ (because $\log e = 1$). Then a one-unit increase in the value of the logged variable is equivalent to multiplying the underlying unlogged variable by the base e, or (rounded) 2.71828. With this in mind, we can interpret b_2 in our regression equation as follows: Our first interpretation that *"an increase of one in the value of log X_2* will (additively) change $\hat{Y}$ by b_2" can be restated as *"multiplying X_2 by e, or 2.71828,* will (additively) change $\hat{Y}$ by b_2."

If b_2 is positive, this change in $\hat{Y}$ is an increase, and if b_2 is negative, it is a decrease. This restated interpretation is certainly an improvement, because it involves X_2 rather than log X_2, and therefore should be much more understandable for readers of our research. However, there may be a practical concern with this standard interpretation. In many instances, multiplication by e—that is, almost tripling the value of X_2—would be an extremely dramatic, perhaps even unrealistic, change or difference in X_2. For example, suppose X_2 is annual income. For income, multiplication by e, or almost tripling the income, would represent a very large change or difference, and this could mislead readers of our research as to the real magnitude of income's effect on $\hat{Y}$. The change in $\hat{Y}$ resulting from such a dramatic change in income might not be the best tool for our assessment of this variable's real-world significance. Of course even without any logging, we sometimes decide to focus our interpretation on something other than a one-unit change in an X, so this is not an unfamiliar issue. It just tends to be highlighted with log X, as a multiplicative increase of e will seem extremely large for many independent variables in many practical research contexts.

Fortunately, it is not difficult to instead consider some smaller multiplicative change in X_2. Suppose, for example, that we decide it would be informative to examine the effect on $\hat{Y}$ of a 25% increase in X_2, which is equivalent to multiplying X_2 by 1.25. With our regression equation taking the form $\hat{Y} = a + b_1 X_1 + b_2 \log X_2$, multiplying X_2 by 1.25 gives us $\hat{Y} = a + b_1 X_1 + b_2 \log (1.25 \times X_2)$. As other X's have been held constant, the change in $\hat{Y}$ is then determined by the change in the part of the regression equation involving X_2, and that has changed from $b_2 \log X_2$ to $b_2 \log (1.25 \times X_2)$.

Using rule #2 for logs [which says that $\log (f \times g) = \log f + \log g$] again, $\log (1.25 \times X_2) = \log 1.25 + \log X_2$, so that the term $b_2 \log (1.25\, X_2)$ can be rewritten as $b_2 \log 1.25 + b_2 \log X_2$. Then the whole equation after multiplying X_2 by 1.25 (that is, increasing X_2 by 25%) is

$$\hat{Y} = a + b_1 X_1 + b_2 \log 1.25 + b_2 \log X_2.$$

If we compare this to how the regression equation looked *before* multiplying X_2 by 1.25 (that was $\hat{Y} = a + b_1 X_1 + b_2 \log X_2$), we notice that the difference in the two equations is $b_2 \log 1.25$. That is, that the effect of multiplying X_2 by 1.25, holding other X's constant, is to change $\hat{Y}$ by $b_2 \log 1.25$. Our calculator shows that $\log 1.25 = 0.223$, so the change in $\hat{Y}$ is $b_2 \times 0.223$. Note that this is the additive effect on $\hat{Y}$ of a multiplicative change in X_2; again our interpretation involves a multiplicative change in the logged variable (in this case X_2), and an additive change in the unlogged variable (in this case Y).

This example involved multiplying X_2 by 1.25, but of course the same principle would apply to any multiplicative change in X_2 that we might consider, such as multiplying X_2 by 1.50 (increasing X_2 by 50%) or multiplying X_2 by 0.90 (decreasing X_2 by 10%). In general, multiplying X_2 by c will change $\hat{Y}$ by $b_2 \log c$; whether this change in $\hat{Y}$ is an increase or decrease will depend on the sign of b_2 and whether c is > 1 (an increase in X_2) or < 1 (a decrease in X_2). For example, multiplying X_2 by 1.10 will change $\hat{Y}$ by $b_2 \log 1.10$, or $b_2 \times 0.095$. See the appendix (Section 5.7) for additional comments on this situation.

To illustrate this with some numbers, suppose that our regression analysis includes three independent variables, with the second and third independent variable logged, and our software estimated the regression equation as

$$\hat{Y} = 22.64 + 2.73 X_1 + 8.22 \log X_2 - 4.47 \log X_3.$$

Logging does not make sense for a dummy variable, so we know that X_2 and X_3 are numerical variables. We will carry out the interpretation of estimated effects for the two logged independent variables; in each case, the interpretation will be based on a multiplicative change in that X. For X_2, multiplying X_2 by e (that is, 2.71828) increases—additively—$\hat{Y}$ by 8.22 units, holding other independent variables constant. Suppose that in the substantive context of this research, this multiplicative change in X_2 seems unreasonably large, and a 50% increase in X_2 seems like it would be more helpful for interpretation. A 50% increase is equivalent to multiplying X_2 by 1.50, so the result of this change in X_2 can be calculated as $b_2 \log c = 8.22 \times \log 1.50$, or $8.22 \times 0.405 = 3.33$. Increasing X_2 by 50% increases $\hat{Y}$ by 3.33 units, holding other X's constant.

b_3 is negative, but the same sort of steps apply, and an increase in X_3 will lead to a decrease rather than increase in $\hat{Y}$. Holding other independent variables constant, multiplying X_3 by e (or 2.71828) decreases $\hat{Y}$ by 4.47 units; again this is an additive change in $\hat{Y}$. As before, we may want to consider some less dramatic change in X_3 in our interpretation. Suppose that a 10% increase in X_3, or multiplication by 1.10, seems appropriate. Multiplying X_3 by 1.10 corresponds to a change in $\hat{Y}$ of $b_3 \log c = -4.47 \times \log 1.10 = -4.47 \times 0.095 = -0.42$, or a decrease of 0.42 units.

Although in most analyses we will be concerned that multiplication by e is too dramatic a change in X for a reasonable interpretation, there could also be instances in which we would like to explore the effect of an even greater multiplicative change in X. Suppose that the value of X_3 quadruples, or in other words is multiplied by 4. The resulting change in $\hat{Y}$ is $-4.47 \times \log 4 = -4.47 \times 1.386 = -6.20$, or a 6.20 unit decrease.

In evaluating real-world significance, the unlogged Y means that the Y-side of the interpretation is the same as usual. On the X-side, the logged X means that we need to be sure that we are considering an appropriately-sized multiplicative change in X when examining the resulting change in $\hat{Y}$. The choice of a multiplicative change in X may not rely much on the sample values of X, as an analyst may decide to consider a 1% or 10% increase in X without referring to most of X's descriptive statistics. In some cases the sample range of X may be helpful in this decision, but the range needs to be thought of in multiplicative terms (for example, the largest value of X in the sample is 7.3 times the smallest value).

The fact that at least one X variable is logged does not create any particular difficulty in calculating the predicted value of Y for some set of X values, as long as we correctly log where needed. Using the regression equation for the preceding example, suppose we want to calculate $\hat{Y}$ when $X_1 = 11$, $X_2 = 25$, and $X_3 = 42$. Then

$$\begin{aligned}\hat{Y} &= 22.64 + 2.73X_1 + 8.22 \log X_2 - 4.47 \log X_3 \\ &= 22.64 + 2.73(11) + 8.22(\log 25) - 4.47(\log 42) \\ &= 22.64 + 30.03 + (8.22 \times 3.219) - (4.47 \times 3.738) \\ &= 62.42.\end{aligned}$$

Again we should be careful not to use a series of such calculations to falsely determine the effect of an additive change in an X that had been logged in the regression. Logging an X always requires that we focus on multiplicative changes in that variable when making any general statements of how it affects $\hat{Y}$.

5.3.3 X and Y are Both Logged

The third scenario, interpretation of b when the X and Y are both logged, essentially combines the main ideas from the first two scenarios above. With both X and Y logged, changes in either variable must be expressed in multiplicative terms; a multiplicative change in X produces a multiplicative change in $\hat{Y}$. In this scenario, the regression equation (here with two X's) is of the form $\widehat{\log Y} = a + b_1 \log X_1 + b_2 X_2$, with Y logged and X_1 logged also. To understand the interpretation of b_1, it is helpful to again use the antilog to express the regression equation in terms of $\hat{Y}$ rather than $\widehat{\log Y}$. The antilog gives $\hat{Y} = e^{a+b_1\log X_1+b_2X_2}$. Applying rule #4 for exponents, this can be further rewritten as $\hat{Y} = e^a\, e^{b_1\log X_1}\, e^{b_2X_2}$. If we increase the value of log X_1 by one unit, the regression equation would become $\hat{Y} = e^a\, e^{b_1(\log X_1+1)}\, e^{b_2X_2} = e^a\, e^{b_1\log X_1+b_1}\, e^{b_2X_2}$. Again applying rule #4, we have $\hat{Y} = e^a\, e^{b_1\log X_1}\, e^{b_1}\, e^{b_2X_2}$.

Notice that this is not quite what we saw for the first scenario (Y is logged, X is not), due to the log X_1 term. But despite that difference, the main form of our interpretation looks similar: increasing log X_1 by 1 will multiply $\hat{Y}$ by e^{b_1}. And as in the second scenario (X is logged, Y is not), we want to discuss changes in X_1 rather than in log X_1, so we note that increasing log X_1 by 1 is equivalent to multiplying X_1 by e. Putting these two pieces together, we can say that multiplying X_1 by e will multiply $\hat{Y}$ by e^{b_1}, holding other independent variables constant. This will be our core interpretation when both X and Y are logged.

As before, it may be helpful to consider some different multiplicative change in X (rather than multiplying it by e) when interpreting this b. Suppose that we multiply X_1 by some value c. After this multiplication, the regression equation becomes $\hat{Y} = e^a\, e^{b_1\log(c\,X_1)}\, e^{b_2X_2} = e^a\, e^{b_1(\log X_1+\log c)}\, e^{b_2X_2} = e^a\, e^{b_1\log X_1+b_1\log c}\, e^{b_2X_2} = e^a\, e^{b_1\log X_1}\, e^{b_1\log c}\, e^{b_2X_2}$. Comparing this to the original equation before multiplying X_1 by c ($\hat{Y} = e^a\, e^{b_1\log X_1}\, e^{b_2X_2}$), it is clear that the change in $\hat{Y}$ is given by the new additional term $e^{b_1\log c}$, so multiplying X_1 by c will multiply $\hat{Y}$ by $e^{b_1\log c}$, holding the other independent variables constant.

It is straightforward to calculate $e^{b_1\log c}$ in steps: first calculate log c, then multiply that by b_1, and finally calculate that power of e. But there is also another quicker way to find this value. This approach takes advantage of some of the rules for logs and exponents discussed above. First, we note that $b_1 \log c = \log c^{b_1}$, according to the rule #3 for logs above, so $e^{b_1\log c} = e^{\log c^{b_1}}$. Next, we remember that the antilog has the form of e with a log in the exponent, so this is the antilog of $\log c^{b_1}$. This means that $e^{\log c^{b_1}} = c^{b_1}$. So our conclusion is that, holding other independent variables constant, multiplying X_1 by c will multiply $\hat{Y}$ by c^{b_1}. If our calculator can do such an operation, this will be somewhat quicker work than calculating $e^{b_1\log c}$. But both approaches to the calculation will lead to the same result (except perhaps for a minor discrepancy due to rounding), so choosing one or the other is a matter of preference. The appendix (Section 5.7) includes additional discussion of the interpretation of b in this scenario.

As an example with some numbers, suppose that the regression equation from our analysis is

$$\widehat{\log Y} = 4.07 + 0.88 \log X_1 - 1.25 \log X_2 + 0.24 X_3.$$

In considering b_1, note that both X_1 and Y have been logged, so changes in both variables will be expressed in multiplicative terms when we make the interpretation. Our standard interpretation is that multiplying X_1 by e will multiply $\hat{Y}$ by $e^{b_1} = e^{0.88} = 2.411$, holding the other independent variables constant. Suppose that we wanted to examine a less dramatic change in X_1, such as a 10% increase, corresponding to multiplying X_1 by 1.10. Calculating $e^{b_1 \log c}$ in steps, c = 1.10 and log c = 0.095, and $e^{b_1 \log c} = e^{0.88 \times 0.095} = e^{0.084} = 1.088$. This means that multiplying X_1 by 1.10 multiplies $\hat{Y}$ by 1.088; that is, a 10% increase in X_1 is associated with an 8.8% increase in $\hat{Y}$. Alternatively, we could calculate c^{b_1}, because $e^{\log c^{b_1}} = c^{b_1}$ as explained in the previous paragraph. $c^{b_1} = 1.10^{0.88} = 1.087$, which is the same as the figure found by calculating $e^{b_1 \log c}$ in steps, except for a small difference due to rounding.

b_2, for logged variable X_2, involves similar calculations and interpretation, differing only in its negative value. Holding other independent variables constant, multiplying X_2 by e will multiply $\hat{Y}$ by $e^{b_2} = e^{-1.25} = 0.287$. For a less dramatic 5% increase in X_2 (which means multiplying X_2 by c = 1.05, so log c = 0.049), we can calculate $e^{b_2 \log c}$ as $e^{-1.25 \times 0.049} = e^{-0.061} = 0.941$. Therefore a 5% increase in X_2 multiplies $\hat{Y}$ by 0.941, for a 5.9% decrease. The same value can be obtained by $c^{b_2} = 1.05^{-1.25} = 0.941$. Although we did not specify the actual variables involved here, it is likely that this estimate of a 5% increase in X_2 leading to a 5.9% decrease in $\hat{Y}$ would be considered substantial enough to have real-world significance. In most data contexts, a multiplicative change in $\hat{Y}$ that is roughly on par with the corresponding multiplicative change in X would be considered large. Note that we do not present the full interpretation of the effect of X_3 on logged Y ($b_3 = 0.24$) here, but of course that interpretation would follow our earlier discussion of the scenario in which Y is logged and X is not.

Finding a predicted value of Y in this scenario of logged Y and logged X does not present any special difficulties. Using our example regression equation $\widehat{\log Y} = 4.07 + 0.88 \log X_1 - 1.25 \log X_2 + 0.24 X_3$, suppose that we are interested in the predicted value of Y when $X_1 = 83.50$, $X_2 = 42.18$, and $X_3 = 7.00$. Then

$$\begin{aligned}\widehat{\log Y} &= 4.07 + 0.88(\log 83.50) - 1.25(\log 42.18) + 0.24(7.00) \\ &= 4.07 + 0.88(4.425) - 1.25(3.742) + 0.24(7) \\ &= 4.967.\end{aligned}$$

This number is for $\widehat{\log Y}$, so we use the antilog to find $\hat{Y}$ as $e^{4.967} = 143.60$. In the first scenario (Y is logged, X is not), we saw that this $\hat{Y}$ could also be obtained by applying the antilog to the entire equation for predicted log Y and then multiplying the resulting terms to find $\hat{Y}$. That is possible in the current scenario as well, with further simplification possible via steps like those we used to arrive at c^{b_1} above. Although we do not go through those details here, again this will result in the same prediction for Y that we just found by a slightly different approach.

5.3.4 Summary

The following table summarizes the key elements of interpretation for b when using logged variables. Of course all interpretations still include the idea of holding other independent variables constant.

Scenario	Standard Interpretation	Special-Case Interpretation
1. Y is logged, X is not	For numerical X, $\hat{Y}$ is multiplied by e^b as X increases (additively) by 1 unit. For dummy X, $\hat{Y}$ for dummy category is e^b times $\hat{Y}$ for reference category.	Increasing (additively) X by c units multiplies $\hat{Y}$ by $e^{bc} = (e^b)^c$.
2. X is logged, Y is not	$\hat{Y}$ increases (additively) by b as X is multiplied by e (2.718).	Multiplying X by c increases (additively) $\hat{Y}$ by b(log c) units (which is a decrease if b(log c) is negative). For example, multiplying X by 1.10 (increasing X by 10%) increases $\hat{Y}$ by b(log 1.10).
3. Y is logged and X is logged	$\hat{Y}$ is multiplied by e^b as X is multiplied by e (2.718).	Multiplying X by c multiplies $\hat{Y}$ by $e^{b \log c} = c^b$ For example, multiplying X by 1.10 (increasing X by 10%) multiplies $\hat{Y}$ by $e^{b \times \log 1.10}$ or 1.10^b.

5.4 Examples of Logged Variables

A wide variety of variables are at least sometimes logged when they are used in social science research studies. As discussed above, a variable is a candidate for logging if we believe that multiplicative differences are more informative than additive differences, or equivalently that the same additive difference has more importance when it occurs at low rather than at high values of the variable. We further need to be sure that in our sample the variable does not include any zero or negative values. If it does include such values, we might consider some adjustment of the original data, such as treating any individuals with zero income as if their income were $1, but a sensible adjustment will not always be possible.

To illustrate the diversity of variables for which we might consider logging, here are examples of logged variables that were used in a variety of studies reported in the *American Sociological Review*, a prominent academic journal in the field of sociology. These variables involve different units of analysis, from individuals to businesses to cities, states, or countries, but are united by the fact that the researchers using the variable in their study decided that the variable met the criteria above for logging. However, we also included one fake entry in the list, a variable that was not actually an example of logging in a real research study. See if you can spot that one before we point it out.

Income/wages (this is logged very often!)

Per capita gross domestic product

Capital accumulation by business

Number of science students in a country

Number of employees of a business firm

Population

Household size

Age of business

Annual percentage change, stock market indices 1920 to 2010

Per capita energy consumption

Occupational sex composition

Homicide rate

Unemployment rate

We said that one entry in this list is false and did not actually appear as a logged variable in a research study in this journal. The one false entry in this list is the annual percentage change in stock market indices. The immediate problem with logging this variable is that for years in which the stock market goes down, this annual percentage change becomes negative. So this variable is not eligible for logging, even if we believed that its nature otherwise made it a candidate. (We could consider some initial transformation that made all its values positive, with the transformed variable then logged, but that could also be viewed as distorting the information in the variable.)

5.5 Example: Income of Workers in a High-Tech Industry

This example will give us practice in running regression analysis with logged variables and interpreting the results. We will consider all three possible situations: (a) Y is logged but X is not, (b) X is logged but Y is not, and (c) both X and Y are logged. A researcher hypothesizes that a person's education and income are positively associated. He also expects that there are income differences among residency statuses such as U.S. citizen, permanent resident, and work visa holder. Because residency status can limit a person's legal employment opportunities, the researcher expects that those whose status allows more options for legal employment will have, on average, higher salaries. If that is true, salaries should on average be highest for citizens, next highest for permanent residents, and lowest for work visa holders.

To test his hypotheses, he obtained data on income, residency status, and education from 23 randomly selected workers in a high-tech industry. Income is measured as annual income in thousands of dollars, education is measured by years of schooling, and residency status is categorized into three types: "U.S. citizen," "permanent resident," and "work visa holder."

The research hypothesis, units of analysis, measurements of independent and dependent variables, and raw data (which are hypothetical) are as follows:

> *Research hypothesis:* Education is positively associated with income. There are income differences among residency statuses, with incomes highest for U.S. citizens and lowest for work visa holders.
>
> *Units of analysis:* 23 workers in a high-tech industry (individual-level data).

Measurements of independent and dependent variables:

> Y: Annual income in \$1,000s (income).
>
> X_1: Years of schooling (education).
>
> X_2: Residency status: "U.S. citizen," "permanent resident," or "work visa holder" (residency status).

Data:

Person ID	Income in \$1,000s (Y)	Education (X_1)	Residency Status (X_2)
1	87	18	U.S. citizen
2	114	18	U.S. citizen
3	60	18	Permanent resident
4	66	20	Permanent resident

(Continued)

Person ID	Income in $1,000s (Y)	Education (X_1)	Residency Status (X_2)
5	116	23	U.S. citizen
6	48	12	Permanent resident
7	53	17	Work visa holder
8	125	22	Permanent resident
9	70	16	Work visa holder
10	96	16	U.S. citizen
11	88	18	U.S. citizen
12	110	16	U.S. citizen
13	72	16	U.S. citizen
14	80	20	Work visa holder
15	49	16	Work visa holder
16	60	16	Work visa holder
17	74	14	U.S. citizen
18	48	12	Work visa holder
19	90	16	Permanent resident
20	109	19	Permanent resident
21	73	18	Work visa holder
22	130	16	Permanent resident
23	125	19	U.S. citizen

Because residency status is a categorical variable, the researcher needs to create dummy variables to represent it, first selecting a reference category and then creating dummy variables for the remaining categories. As discussed in Chapter 3, the choice of reference category does not influence the results in a fundamental way. After selecting "permanent resident" as the reference category, he created dummy variables for "U.S. citizen" and "work visa holder." The resulting dummy variables are as follows:

Person ID	Residency Status (X_2)	"U.S. Citizen" Dummy (X_{2a})	"Work Visa" Dummy (X_{2b})
1	U.S. citizen	1	0
2	U.S. citizen	1	0
3	Permanent resident	0	0
4	Permanent resident	0	0

Person ID	Residency Status (X_2)	"U.S. Citizen" Dummy (X_{2a})	"Work Visa" Dummy (X_{2b})
5	U.S. citizen	1	0
6	Permanent resident	0	0
7	Work visa holder	0	1
8	Permanent resident	0	0
9	Work visa holder	0	1
10	U.S. citizen	1	0
11	U.S. citizen	1	0
12	U.S. citizen	1	0
13	U.S. citizen	1	0
14	Work visa holder	0	1
15	Work visa holder	0	1
16	Work visa holder	0	1
17	U.S. citizen	1	0
18	Work visa holder	0	1
19	Permanent resident	0	0
20	Permanent resident	0	0
21	Work visa holder	0	1
22	Permanent resident	0	0
23	U.S. citizen	1	0

5.5.1 Income (Y) is Logged, X's are Not

This section discusses the example under the scenario of income (Y) having been logged, interpreting b's for the unlogged independent variable education (X_1) and residency status dummy variables X_{2a} and X_{2b} (which of course are not logged either). The data with log-transformed income, rounded to two decimal places (when statistical software is doing the transformation, it will retain more decimal places), look like this:

Person ID	Income in $1,000s (Y)	Logged $1,000s of Income
1	87	4.47
2	114	4.74
3	60	4.09
4	66	4.19

(Continued)

Person ID	Income in $1,000s (Y)	Logged $1,000s of Income
5	116	4.75
6	48	3.87
7	53	3.97
8	125	4.83
9	70	4.25
10	96	4.56
11	88	4.48
12	110	4.70
13	72	4.28
14	80	4.38
15	49	3.89
16	60	4.09
17	74	4.30
18	48	3.87
19	90	4.50
20	109	4.69
21	73	4.29
22	130	4.87
23	125	4.83

Because multiple X's are included in the model, it is important to first check for collinearity by looking at correlations among the X's, including the dummy variables. For these data, the results of correlation analysis among the X's are as follows:

	Education (X_1)	U.S. Citizen Dummy (X_{2a})	Work Visa Dummy (X_{2b})
Education (X_1)	1.000	0.103	-0.199
U.S. Citizen Dummy (X_{2a})	0.103	1.000	-0.530
Work Visa Dummy (X_{2b})	-0.199	-0.530	1.000

The highest (in absolute value) correlation is –0.530 between the U.S. citizen dummy (X_{2a}) and the work visa dummy (X_{2b}). As explained in Chapter 3, it is often the case that dummies representing categories of the same independent variable show high negative correlations, though here –0.530 is not so high (because the reference category has quite a few respondents in it, one dummy's value is less closely associated with the others'). In any case, there does not seem to be a collinearity issue here.

Here is output for the multiple regression analysis using logged income in $1,000s as the dependent variable:

$R^2 = 0.575$

SSE = 0.993

	b	SE of b	t	p-value
Intercept (constant)	3.442	0.337	10.199	<0.001
Education (X_1)	0.056	0.019	3.042	0.007
U.S. citizen dummy (X_{2a})	0.134	0.115	1.161	0.260
Work visa dummy (X_{2b})	−0.263	0.124	−2.121	0.047

The regression output shows first that R^2 is moderately high (0.575), suggesting reasonable quality of predictions in the sample, but with some important discrepancies too. 57.5% of the variance in logged income (Y) is explained by this set of X's, so the model is likely missing some important X's that might influence income (Y). The output also shows that the effect of education (X_1) on logged income (Y) is statistically significant, controlling for residency status (X_2).

When (as here) a categorical independent variable requires more than one dummy, we will rely on an F-test to make an overall assessment of statistical significance. Full results for the regression that excludes the residency status variables are not shown, but SSE = 1.599 in that analysis. F is then calculated as (shown in Chapter 3)

$$F = \frac{(SSE_0 - SSE_1)/s}{SSE_1/df_1}$$

$$F = \frac{(1.599 - 0.993)/2}{0.993/19} = 5.827$$

The F-table with (2, 19) df is used because s = 2 (the number of dummy variables in the set being tested) and $df_1 = N - (\#\ X\text{'s in the full model}) - 1 = 23 - 3 - 1 = 19$. The table shows that the p-value for F = 5.827 is between 0.010 and 0.025 (more precisely, from statistical software or an online F-calculator, 0.011). This p-value is less than the 0.05 cutoff, so the regression with the two residency status dummies does fit the sample data significantly better than the regression without them, and the researcher decides that there is at least some statistically significant difference in predicted income among the residency statuses (X_2), controlling for education (X_1).

As in Chapter 3, p-values from t-statistics for individual dummy variables indicate whether there is a statistically significant difference in predicted income between the category named in the dummy and the reference category.

The p-value from the t-statistic for the U.S. citizen dummy is 0.260, indicating that the difference in predicted income between U.S. citizens and permanent residents, while holding education (X_1) constant, is not statistically significant. But the p-value for work visa ($p = 0.047$) does indicate a statistically significant difference in predicted income between work visa holders and permanent residents.

To examine statistical significance of the difference between U.S. citizens and work visa holders, the researcher can switch the reference category to either U.S. citizen or work visa and rerun the regression analysis. After choosing work visa as the reference category and rerunning the regression analysis with the U.S. citizen and permanent resident dummies, the p-value for the U.S. citizen dummy is 0.003. This indicates a statistically significant difference in predicted income between U.S. citizens and work visa holders.

The next step is the interpretation of b's. The output shows that b for education (b_1) is 0.056. This is the case of logged Y and a numerical X, so we interpret this as indicating that, holding residency status (X_2) constant, predicted income ($\hat{Y}$) is multiplied by $e^{b_1} = e^{0.056} = 1.058$ as education (X_1) increases by 1 year. In other words, predicted income ($\hat{Y}$) increases by 5.8% for each additional year of schooling, which seems large enough to be a meaningful increase in real-world terms. Suppose we want to compare two workers in this industry (with the same residency status) who differ by 4 years of education, corresponding to the difference between a college degree and a high school diploma. To understand the impact of a 4-year, rather than 1-year, difference in education, we calculate $e^{b_1 \times 4} = e^{0.056 \times 4} = e^{0.224} = 1.251$. This means that the predicted income ($\hat{Y}$) of the worker with 4 more years of education is 1.251 times that of the less educated worker. Alternatively, this can be reported as the more educated worker having a 25.1% greater predicted income. These figures could also be found by calculating 1.058^4, following log rule #5 that $e^{b_1 c} = (e^{b_1})^c$; any minor difference is simply due to rounding.

For a dummy variable X, our interpretation is that $\hat{Y}$ for the named category is e^b times $\hat{Y}$ for the reference category. The b for the U.S. citizen dummy (b_{2a}) is 0.134, so the predicted income ($\hat{Y}$) for U.S. citizens is $e^{b_{2a}} = e^{0.134} = 1.143$ times the predicted income ($\hat{Y}$) for permanent residents, controlling for education (X_1). In other words, the predicted income ($\hat{Y}$) of a U.S. citizen is 14.3% greater than that of a permanent resident, holding education (X_1) constant. However, note that the predicted income difference between a U.S. citizen and a permanent resident is not statistically significant ($p = 0.260$). b for the work visa dummy (b_{2b}) is −0.263, meaning that, holding education (X_1) constant, the predicted income ($\hat{Y}$) for a work visa holder is $e^{b_{2b}} = e^{-0.263} = 0.769$ times the predicted income ($\hat{Y}$) for a permanent resident. That is, the predicted income ($\hat{Y}$) for a work visa holder is 23.1% lower than that for a permanent resident.

Reanalyzing after switching the reference category to work visa focuses attention on the predicted income difference between U.S. citizens and work visa holders. As noted above, this difference is statistically significant ($p = 0.003$). $b = 0.397$, which can also be found from the first analysis as the difference in the b's for the U.S. citizen and work visa dummies: $0.134 - -0.263 = 0.397$. Then the predicted income ($\hat{Y}$) for U.S. citizens is $e^{0.397} = 1.487$ times the predicted income ($\hat{Y}$) for work visa holders; equivalently, the predicted income ($\hat{Y}$) for U.S. citizens is 48.7% higher than that for work visa holders, holding education (X_1) constant. Together these results indicate that, when controlling for education (X_1), predicted income ($\hat{Y}$) is the highest for U.S. citizens, followed by permanent residents and work visa holders, but with no statistically significant difference between U.S. citizens and permanent residents. The statistically significant differences also appear to be large in real terms.

5.5.2 Education (X_1) is Logged, Income (Y) is Not

This example illustrates the scenario in which income (Y) has not been transformed, but education (X_1) has been logged. As noted above, there is usually a good theoretical case for logging income, but for the purpose of this example it has not been logged; of course the dummy variables for residency status (X_{2a} and X_{2b}) cannot be logged. The data with log-transformed education (X_1) look like this, again with rounding to two decimal places:

Person ID	Education	Logged Education (X_1)
1	18	2.89
2	18	2.89
3	18	2.89
4	20	3.00
5	23	3.14
6	12	2.48
7	17	2.83
8	22	3.09
9	16	2.77
10	16	2.77
11	18	2.89
12	16	2.77
13	16	2.77
14	20	3.00
15	16	2.77
16	16	2.77
17	14	2.64
18	12	2.48

(Continued)

Person ID	Education	Logged Education (X_1)
19	16	2.77
20	19	2.94
21	18	2.89
22	16	2.77
23	19	2.94

Correlation analysis for the X's (logged education, U.S. citizen dummy, and work visa dummy) indicates that again the only even moderately high correlation is between the two dummies representing residency status (r = –0.530), so there is little concern about the possibility of collinearity. Here is output for the multiple regression analysis with logged education (X_1):

$$R^2 = 0.536$$

$$SSE = 7{,}302.615$$

	b	SE of b	t	p-value
Intercept (constant)	-122.979	76.343	-1.611	0.124
Logged education (X_1)	74.623	26.658	2.799	0.011
U.S. citizen dummy (X_{2a})	7.824	9.881	0.792	0.438
Work visa dummy (X_{2b})	-23.277	10.606	-2.195	0.041

From R^2, 53.6% of the variance in income is explained by this set of X's, suggesting that there may be other important predictors of income that have not been included in this regression. The output also shows that the effect of logged education (X_1) on income (Y) is statistically significant, controlling for residency status (X_2).

To make an overall assessment of statistical significance for the categorical variable with more than one dummy, we need the sum of squared residuals from a regression omitting this set of dummy variables. Here this was 11,131.775, so F is calculated as

$$F = \frac{(SSE_0 - SSE_1)/s}{SSE_1/df_1}$$

$$F = \frac{(11{,}131.775 - 7{,}302.615)/2}{7{,}302.615/19} = 4.981$$

With (2, 19) df, the F-table shows that the p-value corresponding to F = 4.981 is between 0.010 and 0.025, indicating that there is some statistically significant difference in predicted income among residency statuses while holding logged education (X_1) constant. (From statistical software or an online calculator, the precise p-value is 0.018.) However, the p-value from the t-statistic for the U.S. citizen dummy (0.438) indicates that the difference between U.S. citizens and permanent residents is not statistically significant. The p-value for the work visa dummy (0.041) shows a statistically significant difference in predicted income between work visa holders and permanent residents. By switching the reference category to work visa, and examining the p-value for the U.S. citizen dummy, the researcher finds that the predicted income difference between U.S. citizens and work visa holders is also statistically significant ($p = 0.006$), controlling for logged education (X_1).

The b for logged education (b_1) is 74.623. This is the scenario of a logged X and an untransformed Y, so predicted income in \$1,000s ($\hat{Y}$) increases by $b_1 = 74.623$ (meaning \$74,623, with income measured in \$1,000s) as education (X_1) is multiplied by e (2.71828), holding residency status (X_2) constant. For a variable like education, multiplication by e—that is, almost tripling the value—would be an extremely dramatic increase. In fact, in this sample the ratio of the highest to lowest values of education (X_1) is only 23/12 = 1.92, so the highest is not quite double the lowest. Clearly it would be more informative to consider what happens to predicted income ($\hat{Y}$) when education (X_1) changes by a more realistic multiple. For instance, one could examine the effect on predicted income ($\hat{Y}$) of a 10% increase in education (X_1), which is equivalent to multiplying education (X_1) by 1.10. The change in predicted income in \$1,000s ($\hat{Y}$) when multiplying education (X_1) by 1.10 is then given by $b_1 \log 1.10 = 74.623 \times 0.095 = 7.089$ (or \$7,089). Even with the rather high mean (\$84,480) and wide range (\$48,000 to \$130,000) of income in this sample, this seems like a meaningful increase in income for a 10% increase in education (X_1), so this effect seems to have real-life significance. Increasing education (X_1) by 15%, 20%, or 25% would also be realistic multiples that could be used in this interpretation.

As dummy variables are never logged, and in this scenario Y is not logged, the interpretation of b's from the dummy variables in this example follows the same form that we saw in Chapter 3. According to the output, b for the U.S. citizen dummy is 7.824, indicating that the predicted income ($\hat{Y}$) difference between U.S. citizens and permanent residents is 7.824 (that is, \$7,824). However, note that this difference is not statistically significant. The output also shows that b for the work visa dummy is –23.277, meaning that permanent residents are predicted to earn \$23,277 more than work visa holders, controlling for logged education (X_1). The predicted income ($\hat{Y}$) difference between U.S. citizens and work visa holders can be found by rerunning the analysis with work visa as the reference or by calculating the difference in the b's above (7.824 – –23.277). Either approach shows that the predicted income ($\hat{Y}$) for U.S. citizens is \$31,101 more than that for work visa holders, controlling for logged education (X_1). Overall, the differences between residency statuses seem meaningfully large in light of the descriptive statistics for income.

5.5.3 Education (X_1) and Income (Y) are Both Logged

This example illustrates the third scenario in which both income (Y) and education (X_1) have been logged. Correlations among the X's are exactly the same as in the previous example of logged X and untransformed Y, so there are no collinearity concerns. The multiple regression output with logged education (X_1) and logged income (Y) looks like this:

$R^2 = 0.585$

$SSE = 0.969$

	b	SE of b	t	p-value
Intercept (constant)	1.672	0.879	1.901	0.073
Logged education (X_1)	0.969	0.307	3.157	0.005
U.S. citizen dummy (X_{2a})	0.127	0.114	1.114	0.279
Work visa dummy (X_{2b})	−0.268	0.122	−2.195	0.041

R^2 indicates that 58.5% of the variance in logged income (Y) is explained by this set of X's. Note that this is a bit higher than in the first example, with income (Y) logged but education (X_1) not, so the predictions of logged income ($\hat{Y}$) are slightly better in general when education (X_1) is also logged. The output also shows that the effect of logged education (X_1) on logged income (Y) is statistically significant, controlling for residency status (X_2).

To make an overall assessment of statistical significance for residency status (X_2), Model 0 omits the residency status dummies, and F is calculated as

$$F = \frac{(SSE_0 - SSE_1)/s}{SSE_1/df_1}$$

$$F = \frac{(1.572 - 0.969)/2}{0.969/19} = 5.912$$

With (2, 19) df, the F-table gives a p-value for F = 5.912 between 0.010 and 0.025 (the precise p-value from statistical software or an online calculator rounds to 0.010), showing that there is at least some statistically significant difference in predicted income among residency statuses. The difference between U.S. citizens and permanent residents is not statistically significant (p-value for the U.S. citizen dummy = 0.279), but there are statistically significant income differences between work visa holders and permanent residents (p-value for the work visa dummy = 0.041), and also between U.S. citizens and work visa holders (p-value for the U.S. citizen dummy = 0.003 when rerunning the regression with work visa as the reference category), controlling for logged education (X_1).

With both education (X_1) and income (Y) logged, changes in either variable must be expressed in multiplicative terms: a multiplicative change in education (X_1) produces a multiplicative change in the predicted income (Y). The b for logged education (X_1) is 0.969, so we can say that multiplying education (X_1) by e will multiply the predicted income ($\hat{Y}$) by $e^{b_1} = e^{0.969} = 2.635$, holding other independent variables constant. Again, it is probably better to consider a less dramatic change in education (X_1), such as the 10% increase used before, corresponding to multiplying education X_1 by 1.10. Then the predicted income ($\hat{Y}$) is multiplied by $e^{b_1 \log 1.1} = e^{0.969 \times 0.095} = e^{0.092} = 1.096$ as education is multiplied by

1.10. That is, the predicted income ($\hat{Y}$) increases by 9.6% as education increases by 10%, holding residency status (X_2) constant. This seems like a meaningfully large effect on income. Alternatively, this can be calculated from $1.10^{0.969} = 1.097$ (with the small difference due to rounding).

The U.S. citizen (X_{2a}) and work visa (X_{2b}) dummy variables are of course not logged, and this aspect of the analysis is like that with logged Y and dummy independent variables before. For interpretation, $\hat{Y}$ for the named category is e^b times $\hat{Y}$ for the reference category. The b for U.S. citizen (X_{2a}) is 0.127, so the predicted income ($\hat{Y}$) for U.S. citizens is $e^{b_{2a}} = e^{0.127} = 1.135$ times the predicted income (Y) for permanent residents, holding (logged) education (X_1) constant. Alternatively, the predicted income ($\hat{Y}$) for U.S. citizens is 13.5% higher than that of permanent residents. However, as before, the difference between U.S. citizens and permanent residents is not statistically significant ($p = 0.279$). The b for work visa (b_{2b}) is –0.268, meaning that the predicted income ($\hat{Y}$) for work visa holders is $e^{b_{2b}} = e^{-0.268} = 0.765$ times the predicted income ($\hat{Y}$) for permanent residents. This can also be reported as the predicted income ($\hat{Y}$) of work visa holders being 23.5% lower than that of permanent residents, controlling for (logged) education (X_1).

As in previous examples, switching the reference category to work visa and rerunning the analysis, or looking at the difference in the two original dummy's b's, will allow the researcher to find the predicted income difference between U.S. citizens and work visa holders. The b for U.S. citizen in the rerunning of the regression, or the difference (0.127 – (–0.268)), is 0.395. The predicted income ($\hat{Y}$) for U.S. citizens is therefore $e^{0.395} = 1.484$ times the predicted income ($\hat{Y}$) for work visa holders; equivalently, predicted income ($\hat{Y}$) for U.S. citizens is 48.4% higher than for work visa holders, holding (logged) education (X_1) constant. The multiplicative differences in income among residency status categories clearly seem large enough to have real-world significance.

5.6 Exercises

5.6.1 Exercise 1: Calculator Work With Logs

This exercise gives practice on interpreting b when using logs; it requires a calculator that can find logs (natural logs) and evaluate exponents. Throughout the exercise we will assume that all relevant p-values were small (< 0.05) and that we could reasonably go on to interpret the b's.

1. Suppose you did a regression with log Y as the dependent variable, and obtained the estimated equation $\widehat{\log Y} = 1.03 + 0.08X_1 - 0.15X_2$.

   (a) What is the multiplicative effect of a one-unit increase in X_1 on $\hat{Y}$? What is it for X_2? (Also express these in terms of percentage changes.) Why do these numbers make sense with respect to the signs of the b's in the equation for the predicted log Y?

   (b) How would you express the results if X_1 and X_2 are dummy variables?

   (c) Find the predicted value of Y for $X_1 = 8$ and $X_2 = 12$.

(d) What is the additive effect of a one-unit increase in X_1 on Y? (Trick question! Why is it a trick question?)

2. Suppose you had the regression $\hat{Y} = 8.1 + 85.32 \log X_1 + 1.24X_2$.

   (a) The 85.32 represents the impact on $\hat{Y}$ of what multiplicative change in X_1?

   (b) If the value of X_1 increases to 1.3 times its previous value (holding X_2 constant), how is $\hat{Y}$ affected? What if X_1 became 25% smaller?

   (c) What is $\hat{Y}$ for $X_1 = 4$ and $X_2 = 3$?

3. Suppose you had the regression $\widehat{\log Y} = 2.03 - 1.1 \log X_1 + 0.83 \log X_2$.

   (a) Find (and express) the effect of X_1 on Y. What about the effect of X_2? (Your answers should be given in terms of X_1 and X_2, not the logs of these two variables.)

   (b) If X_1 is tripled (holding X_2 constant), what happens to $\hat{Y}$?

   (c) What is $\hat{Y}$ for $X_1 = 15$ and $X_2 = 46$?

5.6.2 Exercise 2: Infant Mortality in 20 Nations

A researcher is interested in what influences a nation's infant mortality rate. She hypothesizes that greater wealth and resource availability decrease the infant mortality rate and also believes that there are regional differences in infant mortality. To examine her hypotheses in a multivariate context, she obtained data on 20 nations for 2016 from the World Bank. Economic development was measured for each nation by GDP per capita in $1,000s (current U.S. dollars). Geographical region included three categories: "South America," "Central America," and "Caribbean." Infant mortality was measured by number of infant deaths per 1,000 births. The data are shown below.

Data:

ID	Country	Infant Mortality	GDP per Capita in $1,000s	Region
1	Argentina	9.7	12.65	South America
2	Bahamas	6.1	30.26	Caribbean
3	Barbados	11.9	15.89	Caribbean
4	Belize	12.8	4.96	Central America
5	Bolivia	29.0	3.12	South America
6	Brazil	14.6	8.64	South America
7	Chile	6.6	13.96	South America
8	Colombia	13.1	5.76	South America

ID	Country	Infant Mortality	GDP per Capita in $1,000s	Region
9	Costa Rica	8.0	11.73	Central America
10	Dominican Republic	25.6	6.79	Caribbean
11	Ecuador	12.7	6.02	South America
12	El Salvador	12.9	3.77	Central America
13	Haiti	55.0	0.73	Caribbean
14	Honduras	16.2	2.38	Central America
15	Jamaica	13.5	4.88	Caribbean
16	Nicaragua	15.3	2.14	Central America
17	Panama	14.3	14.33	Central America
18	Paraguay	18.4	4.08	South America
19	Peru	12.1	6.03	South America
20	Uruguay	7.3	15.30	South America

1. Log infant mortality rate, and use that as the dependent variable in a multiple regression analysis with GDP per capita and region as independent variables. Interpret the results as fully as possible.
2. Log GDP per capita, and use that as an independent variable in a multiple regression analysis of (unlogged) infant mortality rates. Interpret the results as fully as possible.
3. Log both GDP per capita and infant mortality (the dependent variable), and use those variables in a multiple regression analysis. Interpret the results as fully as possible.
4. The researcher attempted to log geographical region. Explain what is wrong with her attempt.
5. Suppose that instead of GDP per capita, the researcher found additional data and decided to calculate GDP per capita growth as (GDP per capita in 2016 – GDP per capita in 2015)/(GDP per capita in 2015).

   Before including this as an independent variable in the analysis, the researcher also considered logging the variable. Is it appropriate for the researcher to log this variable? Explain.

5.7 Appendix

We mentioned that one of the properties of the natural log is that if d is a small number (close to zero), then log (1 + d) will approximately equal d. Using the antilog, we can also say that e^d is approximately (1 + d), because $e^{\log(1+d)} = (1 + d)$. For example, 1.06 can be written as (1 + 0.06), and a calculator shows that the natural log of 1.06 is 0.058, which is very close to the 0.06 that would be suggested by this rule of thumb. Also, on the calculator $e^{0.06} = 1.062$, which is very close to 1 + 0.06. This also works for small negative numbers. The calculator tells us that $e^{-0.104} = 0.901$, and the approximation would be very close: (1 + –0.104) = 0.896. The quality of the approximation declines as d gets further from zero; for instance $e^{0.40} = 1.492$, so that the approximation 1.40 is not very accurate. But when d is close to zero, the approximation is quite successful.

This idea can be used for interpretation involving log X, especially for fields in which it is traditional to focus on a 1% increase in X. Suppose that Y is not logged, and we are considering a 1% increase in X. The discussion in this chapter shows that $\hat{Y}$ would then change by b × log 1.01. Remembering our rule of thumb about log (1 + d) for a small number d, we know that log 1.01 will be quite close to 0.01. In fact, when rounding to three decimal places, the calculator says that log 1.01 = 0.010; because d in this case is so close to zero, the approximation is very good. Then it is easy to remember that for a 1% increase in X, $\hat{Y}$ will change by (b × 0.010).

Now suppose that Y has also been logged, so we are in the logged Y and logged X scenario. For a 1% increase in X, the multiplicative effect on Y will be $e^{b \log(1.01)}$, which from the earlier discussion is $e^{(b \times 0.01)}$. As long as b is not very large, (b × 0.01) will be pretty small, so $e^{(b \times 0.01)}$ will be approximately (1 + [b × 0.01]). To express this multiplicative effect in percentage terms, as usual we would look at the difference from 1, multiplied by 100 to make a percentage. The difference from 1 is (b × 0.01), and (b × 0.01) × 100 = b. So in this logged Y and logged X situation, b is approximately the percentage change in $\hat{Y}$ for a 1% increase in X. Unless b is unusually large (in either the positive or negative direction), this approximation will be quite good.

In the main text, $b_1 = 0.88$ and $b_2 = -1.25$ were two of the values that we used to illustrate the logged Y and logged X scenario. For a 1% increase in X, our calculator work would show that these b's imply that $\hat{Y}$ would be multiplied by 1.0088 and 0.9876, respectively. These multiplicative effects represent a 0.88% increase and a 1.24% decrease in $\hat{Y}$, respectively, which in each case is exactly or almost exactly a change of b % in $\hat{Y}$. So unless b is uncommonly large, in the logged Y and logged X situation we can look at b and immediately interpret this as the percentage change in $\hat{Y}$ when X increases by 1%, at least to a close approximation.

6

Nonlinear Relationships in Multiple Regression

The straight-line relationship between X and Y is one of the main features that we grasp when first encountering simple regression. And even when we extend our view to multiple regression, we continue to think about straight-line relationships between an X and Y while controlling for the other X's in the regression. However, the linear regression framework is quite flexible and actually can accommodate relationships between variables that do not follow this straight-line form. We saw this first with dummy variables, as the categorical information represented by dummy variables does not lend itself to the usual image of a straight-line relationship in a scatterplot. But even when an independent variable is numerical rather than categorical, we can use multiple regression to investigate relationships between X and Y that are *nonlinear* (that is, not represented by a straight line) rather than linear. One example of this is the use of logged variables as discussed in Chapter 5. The motivation that we presented for logging relied on the idea that a given additive change in some variable could have different meaning depending on whether it occurred in the context of a high or low value for the variable. However, as we show in this chapter, logging can also be motivated by a desire to represent a particular kind of nonlinear relationship between an independent and dependent variable.

The key to representing a nonlinear relationship lies in transforming a variable or variables so that when the transformed variables are used in a multiple regression, the straight-line relationships involving the transformed variable(s) are in fact expressing nonlinear relationships for the original, untransformed variable(s). The log transformation in Chapter 5 is one example of this approach, and we begin the current chapter by looking at the log more closely. The rest of the chapter then explores some other common transformations of variables that allow a multiple regression analysis to express nonlinear relationships between variables. There are many more possibilities beyond those discussed here, but these common transformations help illustrate the potential of this important extension of multiple regression. A full example will illustrate how to incorporate a nonlinear relationship into multiple regression and interpret the results. The chapter concludes with exercises.

6.1 Nonlinear Relationships Implied by Logs

The discussion of logs in Chapter 5 emphasized thinking about whether multiplicative or additive differences better capture a variable's nature as a main basis for deciding whether or not to log that variable. But a somewhat different view of logging would emphasize the type of nonlinear relationship between X and Y that is implied when one or both are logged in our regression. The nonlinear relationship between X and Y implied by logs also gives another perspective on our warning about not trying to report a general finding regarding additive changes in a variable that was logged in our analysis. Here we explore this nonlinear relationship in the context of simple rather than multiple regression, but the ideas carry over to multiple regression too. As before, we will assume that any variable that we decide to log does not take on any zero or negative values in our sample (remembering rule #1 for logs and exponents in Chapter 5).

Suppose that we decided that multiplicative thinking should apply to Y, but not X, so that our regression included log Y and the original, unlogged X ($\widehat{\log Y} = a + bX$). Further suppose that our analysis resulted in the estimated regression $\widehat{\log Y} = 3.472 + 0.438X$. This is a straight-line model for the relationship between $\widehat{\log Y}$ and X, and, as we saw in the previous chapter, it implies that (additive) changes in X produce multiplicative changes in $\hat{Y}$. To better understand this relationship between X and Y, we can plug in some X values and see the corresponding $\widehat{\log Y}$ and $\hat{Y}$; for convenience, we will simply plug in several positive integers (1, 2, 3, 4, and 5 in this case) as X values, though of course we could use any X values in the equation.

X Value	$\widehat{\log Y} = 3.472 + 0.438X$	$\hat{Y} = e^{\widehat{\log Y}}$
1	3.472 + 0.438 (1) = 3.910	$e^{3.910} = 49.899$
2	3.472 + 0.438 (2) = 4.348	$e^{4.348} = 77.324$
3	3.472 + 0.438 (3) = 4.786	$e^{4.786} = 119.821$
4	3.472 + 0.438 (4) = 5.224	$e^{5.224} = 185.675$
5	3.472 + 0.438 (5) = 5.662	$e^{5.662} = 287.724$

The table shows that, as we knew would be the case, $\widehat{\log Y}$ increases by $b = 0.438$ for any one-unit increase in X. For example, $\widehat{\log Y}$ increases from 3.910 to 4.348 as X increases from 1 to 2, and from 4.348 to 4.786 as X increases from 2 to 3. These and the subsequent steps in the table represent increases of 0.438

in $\widehat{\log Y}$. Recalling the scenario of a logged Y and an unlogged X from Chapter 5, we also knew that a one-unit increase in X would multiply the predicted by Y by e^b; here $e^{0.438} = 1.550$. This is borne out in the table, as we can confirm that when we divide the $\hat{Y}$ shown for an X by the $\hat{Y}$ shown for the previous value of X, we always get 1.550. That is, (77.324/49.899) = (119.821/77.324) = (185.675/119.821) = (287.724/185.675) = 1.550. In each instance, increasing X by one unit multiplied $\hat{Y}$ by 1.550.

As we warned in Chapter 5, however, a one-unit increase in X does not have a consistent additive effect on $\hat{Y}$. The additive change in $\hat{Y}$ was 77.324 – 49.899 = 27.425 when X increased from 1 to 2, 119.821 – 77.324 = 42.497 when X increased from 2 to 3, 185.675 – 119.821 = 65.854 when X increased from 3 to 4, and 287.724 – 185.675 = 102.049 when X increased from 4 to 5. The additive change in $\hat{Y}$ implied by a one-unit increase in X therefore depends on what X value we are starting from in the calculation. The additive increase in $\hat{Y}$ when X goes from 2 to 3 is greater than the additive increase when X goes from 1 to 2, and so on. The positive b (0.438), and the $e^b > 1$ (1.550), told us that there is a positive relationship between X and Y, so that $\hat{Y}$ becomes greater as X increases. But the table also shows that $\hat{Y}$ is increasing *at an increasing rate* as X increases. This is consistent with the motivation for logging Y presented in Chapter 5, in which we logged when we believed that the same additive difference in Y is more meaningful at its lower than at its higher values. If so, then the additive increment to $\hat{Y}$ when X increases by one unit needs to be greater when it occurs in the context of bigger values of $\hat{Y}$, because the same additive increment would not be as meaningful.

What if b had been negative? Say that our estimated regression was $\widehat{\log Y} = 4.857 - 0.284\,X$. We can construct a similar table for this regression:

X Value	$\widehat{\log Y} = 4.857 - 0.284\,X$	$\hat{Y} = e^{\widehat{\log Y}}$
1	4.857 – 0.284 (1) = 4.573	$e^{4.573} = 96.834$
2	4.857 – 0.284 (2) = 4.289	$e^{4.289} = 72.894$
3	4.857 – 0.284 (3) = 4.005	$e^{4.005} = 54.872$
4	4.857 – 0.284 (4) = 3.721	$e^{3.721} = 41.306$
5	4.857 – 0.284 (5) = 3.437	$e^{3.437} = 31.094$

As before, there is a consistent multiplicative effect of a one-unit increase in X, given by $e^b = e^{-0.284} = 0.753$. A one-unit increase in X decreases $\hat{Y}$ by 24.7% (or multiplies $\hat{Y}$ by 0.753), no matter what X values are involved (for instance, 72.894/96.834 = 0.753). However, again the additive effect on Y is not consistent. In additive terms, $\hat{Y}$ changed by 72.894 – 96.834 = –23.94 when X increased from 1 to 2, 54.872 – 72.894 = –18.022 when X increased from 2 to 3, 41.306 – 54.872 = –13.566 when X increased from 3 to 4, and 31.094 – 41.306 = –10.212 when X increased from 4 to 5. Looking at differences between the $\hat{Y}$ values in the table, the additive change in $\hat{Y}$ implied by a one-unit increase in X is decreasing in magnitude (that is, getting closer to zero) as the value of X increases. $\hat{Y}$ is decreasing *at a decreasing* rate as X increases. Again, this is consistent with the earlier motivation for logging a variable: with the same additive difference being more meaningful at lower values of Y, the additive change (decrease) in $\hat{Y}$ when X increases needs to be greater when it takes place at an X value that is associated with a larger value of $\hat{Y}$.

If we plot $\hat{Y}$ against X for these two regressions, we can see the nonlinearity of the implied relationship between X and Y. Plotting $\widehat{\log Y}$ against X would produce a straight line, but shifting from $\widehat{\log Y}$ to $\hat{Y}$ reveals the nonlinear relationship. For Figure 6.1, the plot of $\widehat{\log Y} = 3.472 + 0.438$ X, we show a range of 0 to 8 for the X values. Of course larger values of X could be plotted too, and when X is a variable for which negative values are possible, those could be used in the plot as well. Because the values of $\hat{Y}$ rapidly become very large as X increases, the two axes cannot realistically have the same scale, and this is reflected in the plot. As indicated in the table above, the plot shows that $\hat{Y}$ is increasing at an increasing rate as X increases. Figure 6.2 is a similar plot of the regression equation $\widehat{\log Y} = 4.857 - 0.284$ X and illustrates that in this case $\hat{Y}$ is decreasing at a decreasing rate as X increases.

Nonlinear relationships between X and Y are also implied under the other log scenarios (X logged but Y not, and both X and Y logged) that we studied in Chapter 5. We illustrate some examples of this sort with the following graphs. We first plot the regression equation for a logged X positively related to an unlogged Y, specifically $\hat{Y} = 18.914 + 3.730 \log X$ (Figure 6.3). Plotted values of X

FIGURE 6.1 ● Plot of $\widehat{\log Y} = 3.472 + 0.438$ X, for X Values From 0 to 8

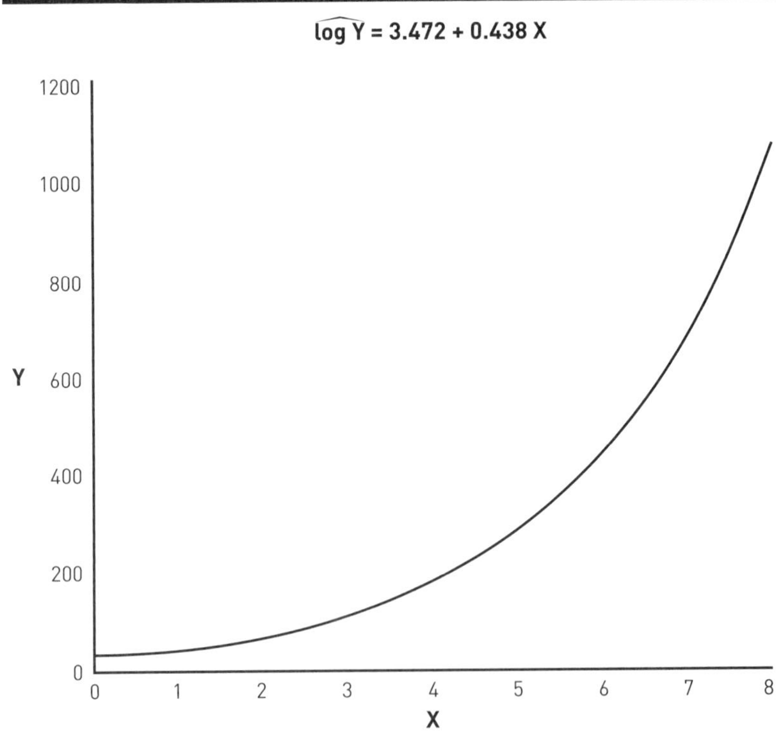

FIGURE 6.2 ● Plot of $\widehat{\log Y} = 4.857 - 0.284\,X$, for X Values From 0 to 8

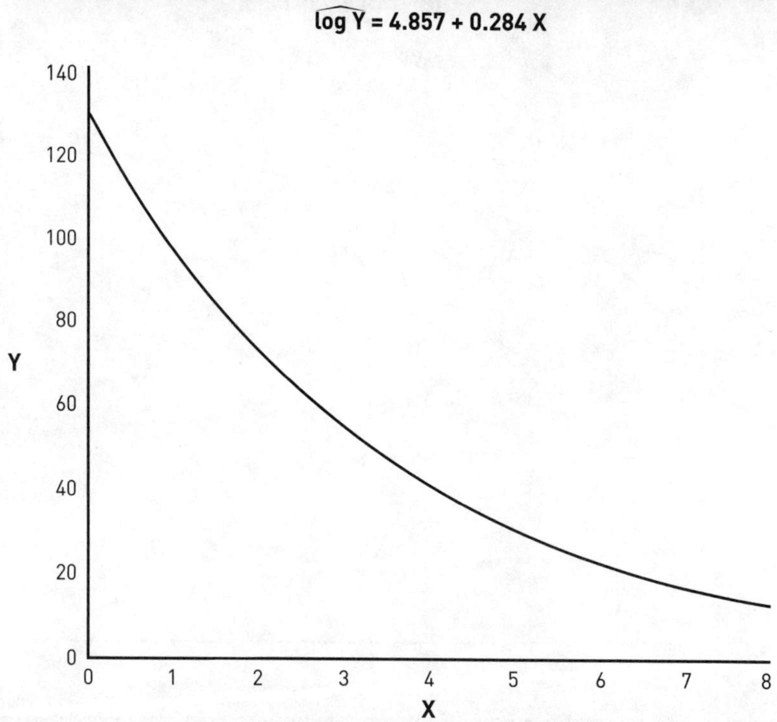

start just above 0 (remember that 0 or negative values cannot be logged) and end at 50. In this case X and Y can reasonably be plotted using the same scale on the axes, and the plot shows that $\hat{Y}$ is increasing as X increases, but at a *decreasing* rate. Notice that for the range of X from 20 or so up to 50, it is pretty difficult to tell whether there is a curve or a slowly increasing straight line. More generally, it becomes harder to distinguish such curves from straight-line relationships when analyzing samples that exhibit a rather restricted range of X values. We do not show a similar plot for a negative relationship, but that graph would show that $\hat{Y}$ is decreasing as X increases, again at a decreasing rate.

Finally, we plot an example of a regression in which both X and Y are logged, with a negative relationship: $\widehat{\log Y} = 5.067 - 0.433 \log X$. Figure 6.4 shows the plot for X values from just over 0 to 200. The negative relationship between X and $\hat{Y}$ is evident, and the decrease in $\hat{Y}$ slows as X increases. Although this general verbal description sounds quite like that for the plot of $\widehat{\log Y} = 4.857 - 0.284\,X$ in Figure 6.2, the specific shape of the curve differs in the two equations.

These examples show that the log transformation in Chapter 5—motivated there by the nature of particular variables in our analysis and a desire to represent multiplicative differences—can also be viewed as a means to express a nonlinear relationship between X and $\hat{Y}$. In practice, it will often seem more straightforward to decide to log based on the rationale offered in Chapter 5 than from a conscious attempt to produce one of these nonlinear relationships. But the graphs demonstrate that we can use our familiar "straight-line" regression framework to accommodate more complicated kinds of relationships.

FIGURE 6.3 ● Plot of $\hat{Y} = 18.914 + 3.730 \log X$, for X Values up to 50

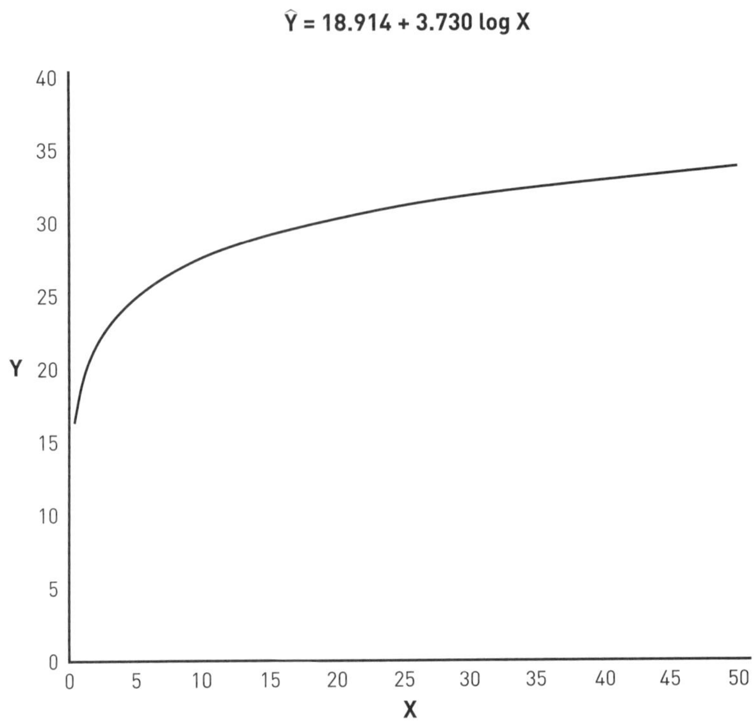

FIGURE 6.4 ● Plot of $\widehat{\log Y} = 5.067 - 0.433 \log X$, for X Values up to 200

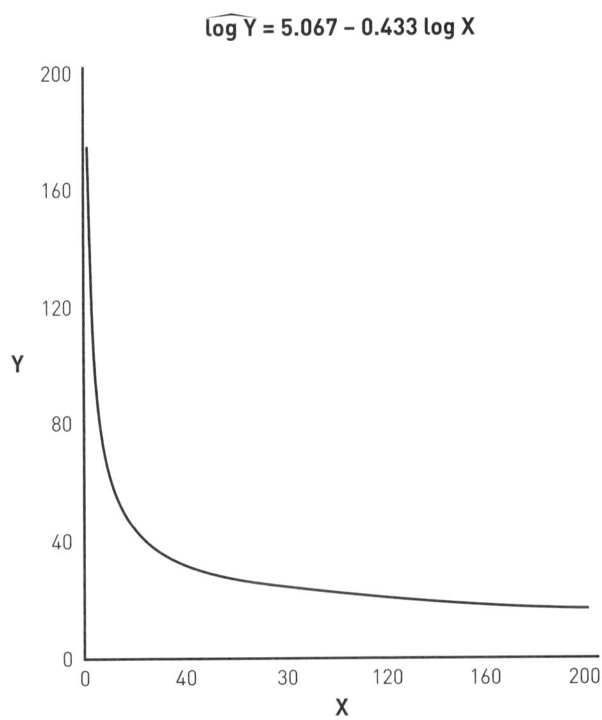

6.2 Squared Independent Variables in Regression

6.2.1 Squared Independent Variables and Curvilinear Relationships

Another common transformation usually does represent an explicit effort to permit a nonlinear relationship between X and Y in our regression. That is the transformation in which we square (multiply by itself) an independent variable. Before looking at the details of the use of this transformation in regression, it is helpful to discuss the motivations that might lead us to this approach. The main appeal of this transformation is that it allows our regression to capture what is often called a *curvilinear* relationship between X and Y, in which the sign of the relationship between X and Y changes from positive to negative, or vice versa, depending on whether we are focusing on the range of X above or below a particular value. That is, such a relationship implies that for one set of values of X, more X is associated with more Y, but for the remaining values of X, more X is associated with less Y.

In graphing the X–Y relationship, one way to express this idea would be a curved line that (loosely speaking) has a "U" shape, or an opposite "upside-down U" shape. In the U-shape, at lower values of X there is a negative relationship with Y, but beyond some value of X the relationship becomes positive. The upside-down U-shape shows the opposite pattern: at lower values of X there is a positive relationship with Y, but beyond some X value the relationship becomes negative. In either image, the slope of the relationship is closer to zero near the X value at which the relationship changes sign and greater in magnitude (a steeper slope) for X values farther away from the point at which the relationship changes sign. Figures 6.5 and 6.6 illustrate the idea of a U-shaped or upside-down U-shaped relationship between X and Y.

As an example of a relationship that might be characterized by this sort of curved line, consider age (X) and delinquent behavior (Y). At relatively young ages, we may expect that the relationship between age and frequency or severity of delinquent behavior is positive. That is, in that younger range, growing older may mean greater involvement with delinquent peers and less with family and

FIGURE 6.5 ● U-Shaped Relationship Between X and Y

U-Shaped Relationship

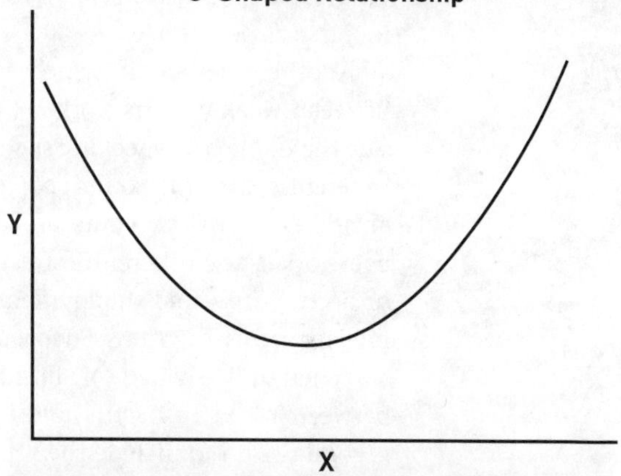

FIGURE 6.6 ● Upside-Down U-Shaped Relationship Between X and Y

Upside-Down U-Shaped Relationship

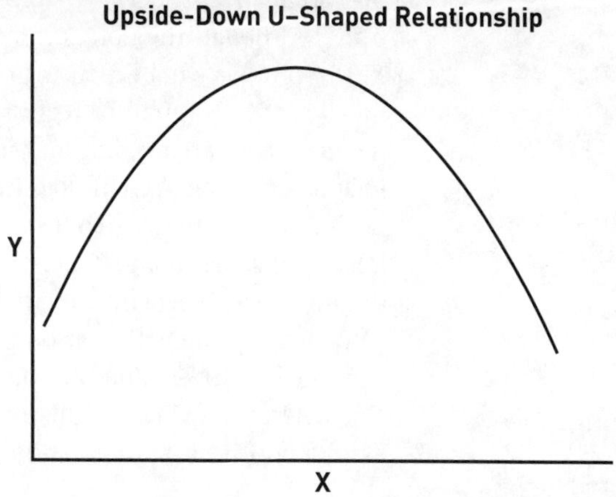

increasing boldness in engaging in behavior that is not approved of by law-abiding adults. However, as youth move into adulthood, age and delinquency likely have a negative relationship, as most individuals grow out of delinquent behavior. Increasing maturity may make such behavior a less attractive source of thrills or fun as one ages, and the potential legal or social costs of engaging in delinquent acts increase with age. This suggests that the relationship between age and delinquency would have the upside-down U-shape described above: a positive relationship between age and delinquency for relatively young ages, but a negative relationship between age and delinquency for older ages.

A similar pattern may hold for the relationship (over the entire life course) between age (X) and physical strength (Y). When focusing on the younger group, physical strength will, on average, increase with age. But beyond some point in life, physical strength will tend to decrease for older people, again suggesting an upside-down U-shape for the relationship. On the other hand, if the physical measurement used for the dependent variable Y is the time required to run or walk 1 mile, then there is likely a U-shaped relationship with age. For young people, increasing age will be associated with shorter run/walk times (as greater speed means a shorter time), but later in life the average time will increase as people age.

Another example in which a U-shape may be expected is the relationship between weekly hours worked (X) and stress (Y) among people who work outside the home or are seeking such work. Unemployment or underemployment is inherently stressful, so for people who are working few (or no) hours per week, an increase in work hours probably results in a reduction of stress. However, overwork is also inherently stressful, and in the range of hours beyond the level of normal full-time employment, increases in work hours are likely accompanied by increasing stress. Such a pattern would indicate a U-shaped relationship.

When a U-shaped or upside-down U-shaped relationship is suspected between Y and an X (still holding other independent variables constant), there is good reason to include the square of that independent variable in the regression. When we decide to include the square of X_1 among the independent variables in the regression, this will normally be in addition to the "plain" X_1, so the regression will include both X_1 and X_1^2, along with whatever other independent variables we have in the regression. If our regression happened to include only one other independent variable (X_2), we would have $\hat{Y} = a + b_1 X_1 + b_2 X_1^2 + b_3 X_2$. Although the notation looks a bit strange with b_3 as the coefficient for X_2, for now we will be focusing on the prior terms on the right-hand side, those involving X_1. Naturally a regression could include more than just one squared independent variable, but for our discussion we will focus on regressions that include only one. As with logs in Chapter 5, it will be easy to create the square of X_1 as a new variable in our statistical software, rather than having to calculate it for each case by hand. Once created, we can include that in the regression just like any other independent variable.

If there is no restriction on the sign of the values of X_1, so that both positive and negative values are possible, then the sign of b_2 (the coefficient on X_1^2) determines whether the implied relationship between X_1 and $\hat{Y}$ has a U- (with a positive b_2) or upside-down U- (with a negative b_2) shape. When X_1 is a variable that only takes on positive values, then we are especially interested in the situation in which b_1 (the coefficient on X_1) and b_2 (the coefficient on X_1^2) have opposite signs, because that will indicate a U- or upside-down U-shaped relationship between X_1 and $\hat{Y}$ within the legitimate (positive) range of X_1. In that case, we can think of b_1 as the main determinant of the relationship for low values of X_1, and b_2 as governing the relationship for high values of X_1. (Below we discuss what "low" and "high" values mean in this situation.) Then a pattern of negative b_1 and positive b_2 will mean a U-shaped relationship, because the relationship is negative for low values of X_1 but positive for high values. On the other hand, a positive b_1 with a negative b_2 will point to an upside-down U-shaped relationship, in which the relationship is positive for low values of X_1 but negative for high values. The pattern of signs of b_1 and b_2 will be a major focus of our interpretation.

6.2.2 Statistical Significance for Squared Independent Variables

We need to think carefully about statistical significance when our regression includes squared independent variables, as the way that we approach this may depend on our larger research context. In many cases, we are exploring the possibility of a curvilinear relationship between X_1 and Y after having already assessed X_1 as being related to Y in a previous regression analysis or having gleaned this from the existing research literature on our topic. In that situation, our main interest probably lies in testing the statistical significance of the coefficient on X_1^2. That refers to the hypothesis that β_2 (the coefficient on X_1^2) = 0, indicating no effect of the squared independent variable X_1^2 in the population. The practical importance of this test is that it helps show whether it is worthwhile for the regression to reflect a more complicated curvilinear relationship between X_1 and $\hat{Y}$. It may be that the simpler straight-line relationship between X_1 and $\hat{Y}$ is actually sufficient to describe our data. If the squared term's coefficient does not significantly differ from 0, we would rather stick with using only the "plain" X_1 in the regression instead of adding the apparently unnecessary complication of the squared term. The statistical significance of X_1^2 is therefore often a main element of our interpretations.

A somewhat different situation is when our use of a squared independent variable in a regression like $\hat{Y} = a + b_1 X_1 + b_2 X_1^2 + b_3 X_2$ is motivated by a particular theory's explicit description of a curvilinear relationship between X_1 and Y, or by past research that found such a relationship. That suggests that we should jointly evaluate the statistical significance of X_1 and X_1^2 by testing the null hypothesis that both of the true coefficients β_1 and β_2 (for X_1 and X_1^2, respectively) are zero. We therefore would test H_0: $\beta_1 = \beta_2 = 0$. If we fail to reject this hypothesis, then there is not good evidence for there being the expected relationship between X_1 and Y. If we instead reject this hypothesis, this suggests a relationship between X_1 and Y, warranting further investigation.

The null hypothesis H_0: $\beta_1 = \beta_2 = 0$ is the same sort of null hypothesis that we set up in Chapter 3 when assessing the statistical significance of a categorical independent variable represented by a set of multiple dummy variables in the regression. In Chapter 3 we used the F-test for that assessment and that same approach can be used here, so this is an instance of the more general uses of the F-test discussed in Appendix 3.8.1. For this F-test we will compare our regression to another that leaves out both X_1 and X_1^2; the regression with these terms is Model 1, and the regression that excludes them is Model 0, with $s = 2$. If the resulting p-value is small (≤ 0.05), we conclude that there is some statistically significant relationship between X_1 and Y, and we can further explore the nature of this relationship with a separate test of statistical significance of the coefficient on X_1^2. That is, we can next test H_0: $\beta_2 = 0$ (in other words, no true effect of X_1^2 on Y) as discussed above, via the usual p-value from the t-statistic for X_1^2. If we do not reject the hypothesis that $\beta_2 = 0$, again we can return to the simpler straight-line relationship between X_1 and Y by excluding X_1^2 from the regression model, as the X_1 term seems to have been driving the result of the F-test. On the other hand, rejecting the hypothesis that $\beta_2 = 0$ calls for the more complicated model with X_1^2.

One wrinkle that arises in practice is that we sometimes find that our software output indicates statistical significance for the X_1^2 term but not for X_1. There is no theoretical prohibition on a zero effect for X_1 existing alongside a nonzero effect for X_1^2 in the population, and we discuss this again when interpreting b_1 and b_2 below. Analysts often end up effectively interpreting a statistically nonsignificant value of b_1 when it occurs alongside a statistically significant b_2, particularly if an F-test for the X_1 and X_1^2 terms together indicates statistical significance.

A related practical complication is that X_1 and X_1^2 may be very highly correlated. When, as in many social science applications, X_1 takes on positive (or zero) values only, then there typically will be quite a high correlation between X_1 and X_1^2, raising the collinearity concern that we have encountered before. In practice it often appears that this high correlation does not require any special action. However, some researchers recommend that, for analysis, the raw values of X_1 should be replaced by their centered form, in which each X_1 value is expressed as a deviation from the sample mean of X_1. (This centering can be done in statistical software, not by hand.) X_1^2 is then constructed from the centered X_1, and the correlation between these new versions of X_1 and X_1^2 will be substantially reduced.

This adjustment is expected to reduce the lower standard error of b_1, while leaving b_2 and its standard error unchanged, and it avoids possible technical problems for software in estimating the coefficients when there is an almost perfect correlation between X_1 and X_1^2. In general, centering X_1 will not otherwise produce fundamentally different results in the analysis, but researchers may be more comfortable interpreting results that are based on the uncentered X_1. This is especially true because the analysis with centered data will produce a different value of b_1, possibly even with a different sign. Even if X_1 is restricted to positive values, its centered version will necessarily have both positive and negative values, as some cases will be above the mean and some below. Then the U- or upside-down U-shape of the estimated relationship between X and Y will not necessarily produce the pattern of signs of b_1 and b_2 discussed above for a positive-valued X, with the shape of the relationship instead determined by the sign of b_2 only. Unless the correlation between the raw X_1 and X_1^2 is extremely close to one, it is probably fine to proceed with analysis of the raw version. (The following text and examples are based on that, but the appendix shows results under centering for the full data example.) Of course it is not difficult to look at the centered version as well; although b_1 and the intercept a produced by the regression output will be different in the two versions, those two sets of estimates are related in a systematic way (as discussed in the appendix), and R^2 will not differ between the two analyses.

6.2.3 Interpreting b's With Squared Independent Variables

With X_1 and X_1^2 in the regression, we need to be careful in interpreting the effect of X_1 on $\hat{Y}$, because we do not interpret b_1 or b_2 in isolation. It does not make sense to think about X_1 increasing by one unit while X_1^2 remains constant, or vice versa. We will still imagine that all other X's are held constant when interpreting X_1's effect, but we need to recognize that a change in X_1 also implies a change in X_1^2, adjusting our standard interpretation in order to consider X_1 and X_1^2 as a set.

When we plot a mathematical function of the form $Y = d + eX + f X^2$, we obtain a curve called a *parabola*, with the sort of U- or upside-down U-shaped form that we discussed above. One feature of a parabola is that there is a particular X value at which the bottom of the U, or the top of the upside-down U, occurs. When we look at a multiple regression like $\hat{Y} = a + b_1 X_1 + b_2 X_1^2 + b_3 X_2$, we can begin our interpretation of the effect of X_1 by finding the value of X_1 at which, holding the other X values constant, the estimated relationship between X_1 and $\hat{Y}$ changes sign (negative to positive as X_1 increases for the U-shaped relationship, or positive to negative for the upside-down U-shaped relationship). It turns out that the value of X_1 at which the relationship changes sign can be calculated as $-b_1/2b_2$, the negative of b_1 divided by twice b_2. The justification for this is not very complicated but does depend on a little calculus, so we will not show it here.

Note that $-b_1/2b_2$ will take a negative value if b_1 and b_2 have the same sign (both positive or both negative). That is, if b_1 and b_2 have the same sign, then the value of X_1 at which the estimated relationship with Y changes sign is < 0. Of course X_1 sometimes will be a variable for which negative values are legitimate, or perhaps X_1 has been centered. Then a negative value for $-b_1/2b_2$ does not raise a special concern. But many of our applications in social science involve independent variables for which only positive or zero values are realistic. If X_1 is such a variable (such as unemployment rate, household income, or number of arrests), then b_1 and b_2 having the same sign implies that the relationship between X_1 and $\hat{Y}$ changes sign at what is, for practical purposes, an impossible value of X_1. If so, then the relationship between X_1 and $\hat{Y}$ effectively does *not* change sign over the possible range of X_1. The inclusion of the squared term for X_1 still means that X_1 and $\hat{Y}$ do not have a straight-line relationship (holding other X's constant), but it is as if we see only part of the U or upside-down U when limiting our attention to realistic (positive) values of X_1.

When X_1 can only take on positive values, then we are probably more interested in the situation of b_1 and b_2 having different signs (one positive and one negative). That situation was discussed earlier in the chapter. In that case $-b_1/2b_2$ will be a positive value, as the leading negative sign in the expression cancels with the negative value of either b_1 or b_2, and the relationship between X_1 and $\hat{Y}$ changes sign at a positive value of X_1. For example, if $b_1 = 8$ and $b_2 = -2$, $-b_1/2b_2$ will be positive (equaling 2). Similarly, if $b_1 = -8$ and b_2 is 2, $-b_1/2b_2$ will also be positive (again equaling 2). Note that this *still* could be an X_1 value that is impossible in practical terms. Suppose that X_1 is age (in years), and when we calculate $-b_1/2b_2$ we get 275. Although this may be a perfectly correct application of the formula by plugging in the b_1 and b_2 that were estimated in our regression, we know that no living population could possibly include a 275-year-old person. If we restrict our focus to realistic values of age (say, up to 100), the estimated relationship between age and Y actually does not change sign. Of course, many variables in our analyses will be less familiar to us than age, so we will not always have a strong sense of what values are or are not realistic. In that case, we can refer to the descriptive statistics for that variable and note whether the calculated $-b_1/2b_2$ is within, or reasonably close to, the range of X_1 values that we observed in our sample. If not, then the X_1 value $-b_1/2b_2$ may not be very realistic in the population from which we sampled, or at least was not within the range of data used to estimate the effect of X_1.

If $-b_1/2b_2$ represents a realistic value of X_1, then the fact that the estimated relationship changes sign at that value of X_1 is an important part of our interpretation of the regression results. For a positively-valued X_1, a negative b_1 and a positive b_2 indicate the U-shaped situation, with a negative relationship between X_1 and $\hat{Y}$ for values of $X_1 < -b_1/2b_2$. In that range, higher values of X_1 are associated with lower values of $\hat{Y}$. However, as X_1 increases, the term involving X_1^2 becomes more dominant in this part of the regression equation, as X_1^2 grows much faster than the plain X_1. For X_1 values $> -b_1/2b_2$, the squared term and its positive coefficient b_2 take over, and in this range higher values of X_1 are associated with higher values of $\hat{Y}$.

If instead b_1 is positive and b_2 is negative, then the results suggest an upside-down U-shaped relationship. For X_1 values $< -b_1/2b_2$, there is a positive relationship: higher values of X_1 are associated with higher values of $\hat{Y}$. Again, though, X_1^2 begins to dominate this part of the regression equation, as it gets larger faster than the plain X_1. So for values of $X_1 > -b_1/2b_2$, there is a negative relationship, with higher values of X_1 associated with lower values of $\hat{Y}$.

Note that if b_1 exactly equaled 0, our calculation of $-b_1/2b_2$ would show that the relationship between X_1 and $\hat{Y}$ changes signs at $X_1 = 0$. Suppose that this occurred with an X_1 (like age, for example) that realistically could only take positive values. In that case, we have the situation mentioned earlier, in which the relationship between X_1 and $\hat{Y}$ would still be nonlinear over the positive values of X_1, but the relationship's sign would never change from positive to negative (or vice versa) over that range. Naturally it is not very likely that our output will show b_1 exactly equal to 0, but this helps us think about how researchers might react to output that showed statistical significance for the X_1^2 term but not for the X_1 term. If b_1 were then treated as it were zero, the interpretation would be of a nonlinear but always positive (or always negative) relationship between X_1 and $\hat{Y}$ in the realistic (positive) range of this X_1. Other researchers, however, might prefer to still include the nonsignificant b_1 in the interpretation of the relationship.

Let us illustrate some of these ideas with an example. Suppose that we have the regression equation $\hat{Y} = 12.35 + 10.28\ X_1 - 0.61\ X_1^2 + 2.73\ X_2$. For this example, again we will not concern ourselves with X_2, but of course in an actual analysis that would also be part of our interpretation of results. Also, we assume that, in assessing statistical significance for X_1, the F-test suggests that adding X_1 and X_1^2 to the regression is worthwhile, and that p-values from t-statistics for both X_1 and X_1^2, or at least for X_1^2, indicate statistical significance. Finally, we assume that X_1 can only take positive values, so we need to consider the signs of both b_1 and b_2 in describing X_1's relationship with Y. We then proceed with interpretation of these b's.

$b_1 = 10.28$ and $b_2 = -0.61$, so there is a positive coefficient on the original X_1 and a negative coefficient on the squared X_1. This pattern indicates an upside-down U-shaped relationship between X_1 and $\hat{Y}$ (holding X_2 constant). For low values of X_1, increasing X_1 increases $\hat{Y}$, as in a positive relationship. For high values of X_1, however, there is a negative relationship, with increases in X_1 associated with decreases in $\hat{Y}$. The value where X_1 changes from "low" to "high" values in this formulation is given by $-b_1/2b_2$. Using the figures here, $-b_1/2b_2 = -[10.28/(2 \times -0.61)]$, which rounds off to 8.43. We therefore estimate that for values of X_1 below 8.43, X_1 and Y are positively related, but the relationship is negative for X_1 values > 8.43. Suppose that descriptive statistics for X_1 showed a range from 2.1 to 15.3 in our sample. Then 8.43 is a realistic value for X_1, and we can say that the upside-down U-shape is an appropriate portrayal of X_1's relationship with $\hat{Y}$ when considered over a range of X_1 values similar to that observed in the sample.

We can illustrate this further by again plugging in some X_1 values and looking at $\hat{Y}$ that we get for each. In this case, we are focusing on only a part of the actual calculation of $\hat{Y}$; by setting the intercept and X_2 aside, we are not using the entire regression equation as needed to really find $\hat{Y}$. But if we assume that X_2 is being held constant while X_1 changes, then changes in this part of the $\hat{Y}$ calculation accurately depict how $\hat{Y}$ changes with X_1, even if here we do not provide enough information to actually calculate specific values of $\hat{Y}$.

X_1 Value	X_1^2 Value	$10.28\ X_1 - 0.61\ X_1^2$	Change in $\hat{Y}$ From 1-Unit Increase in X_1
5	25	10.28 (5) – 0.61 (25) = 36.15	–
6	36	10.28 (6) – 0.61 (36) = 39.72	39.72 – 36.15 = 3.57
7	49	10.28 (7) – 0.61 (49) = 42.07	42.07 – 39.72 = 2.35
8	64	10.28 (8) – 0.61 (64) = 43.20	43.20 – 42.07 = 1.13
9	81	10.28 (9) – 0.61 (81) = 43.11	43.11 – 43.20 = –0.09
10	100	10.28 (10) – 0.61 (100) = 41.80	41.80 – 43.11 = –1.31
11	121	10.28 (11) – 0.61 (121) = 39.27	39.27 – 41.80 = –2.53

Our calculation above indicated 8.43 as the value of X_1 at which the relationship between X_1 and Y changes from positive to negative. In the table, we can see that when X_1 is smaller than this, increases in X_1 lead to increases in $\hat{Y}$ (again we do not have the entire calculation of $\hat{Y}$, but the intercept is constant and we assume that X_2 is not changing). These increases in $\hat{Y}$ flatten out as the values of X_1 get larger, so that the increase in $\hat{Y}$ when X_1 increases from 5 to 6 is greater than when X_1 increases from 6 to 7 (from the table, an increase of 3.57 vs. an increase of 2.35). It is not shown in the table, but the maximum that this portion of the $\hat{Y}$ calculation reaches is 43.31, occurring when $X_1 = 8.43$ (calculated from 10.28 [8.43] – 0.61 [71.06] = 43.31); $\hat{Y}$ still increases slightly as X_1 increases from 8 to 8.43. But at $X_1 = 8.43$, the relationship changes sign to negative, so that beyond that value, increasing X_1 leads to a decrease in $\hat{Y}$. These decreases in $\hat{Y}$ pick up speed as X_1 continues to increase: the decrease in $\hat{Y}$ as X_1 goes from 9 to 10 is smaller in magnitude than the decrease as X_1 goes from 10 to 11 (a decrease of 1.31 vs. a decrease of 2.53), and so on.

Figure 6.7 gives additional perspective, depicting the relationship between X_1 and $\hat{Y}$ as indicated by this regression equation. Again these are not the actual $\hat{Y}$ values themselves, as the Y-axis refers to only the part of the $\hat{Y}$ calculation that involves X_1, not that involving the intercept or X_2, as we treat X_2 as being held constant. The pattern shown in the table and discussed above is clearly visible in the plot. The value of X_1 at which the relationship changes sign—here 8.43—is marked.

This example again reinforces that, as we already saw with the log transformation, we can no longer make the familiar "straight-line" interpretation

FIGURE 6.7 ● Plot of $10.28\,X_1 - 0.61\,X_1^2$.

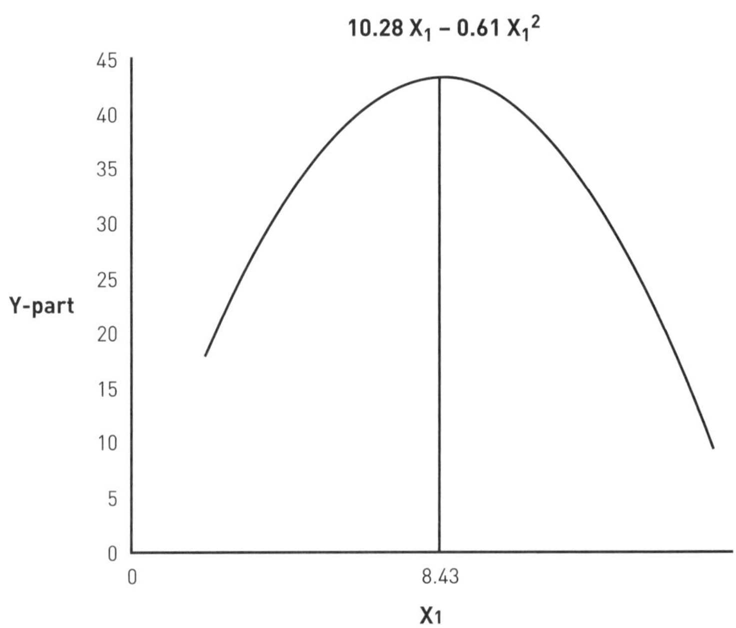

of a one-unit increase in X when considering the b's for squared independent variables. Changes in X_1 will imply changes in X_1^2 as well, so for interpretation X_1 and X_1^2 must be kept together as a package. The discussion, table, and graph above illustrate how the resulting nonlinearity in the relationship prevents us from making a simple statement of how $\hat{Y}$ changes when X_1 increases by one unit. One approach is to abandon "one unit" when considering changes in X_1, and focus on how an infinitesimal change in X_1 would affect $\hat{Y}$ at different values of X_1, via calculus. Interpretation under this approach can also be challenging, and we do not explore it here.

This also complicates our assessment of real-life significance. We need to be a bit more expansive in how we convey results for such a variable when expressing our findings in a paper or report, as we cannot describe the relationship between such an X_1 and Y as compactly as we can in the "straight line" situation. A table like the one here showing how $\hat{Y}$ changes across some list of X_1 values may help us make this assessment. As before, if a one-unit change in X_1 seems too big or small to be ideal in assessing real-life significance, we could look at some different spacing of the X_1 values, as long as those X_1 values were still reasonable in light of the range that we observed in our sample. In any case, we still aim to assess whether appropriate changes in X_1 can result in meaningful changes in $\hat{Y}$, but with more care required when X_1^2 is involved.

6.3 Example: Female Labor Participation in 27 Nations

This example will give us practice in investigating nonlinear relationships in regression analysis and interpreting results. A researcher hypothesizes that a country's economic development and female labor participation have a curvilinear relationship. In a poor country, female labor participation will typically be high, because everyone in a household has to work to meet subsistence needs. In a rich country, female labor participation will likely also be high, reflecting a generally high status of women in society. Female labor participation should be at its lowest in countries with moderate levels of economic development. She also hypothesizes that a country's fertility rate influences its level of female labor participation. Due to cultural differences concerning women's roles and substitution of childrearing for paid work, the fertility rate may be negatively related to female labor participation.

To test her hypotheses, she collected data on female labor participation, economic development, and fertility rate for 27 countries (with Puerto Rico appearing in the data as a distinct country) in 2016 from the World Bank. Female labor participation is measured as percentage of females who are in the labor force, calculated as number of females in the labor force divided by the total female population over age 15, times 100. Economic development is indicated by GDP per capita, in $1,000s. Fertility rate is measured as the average number of children per adult female.

The research hypothesis, units of analysis, measurements of independent and dependent variables, and raw data are as follows:

Research hypothesis: Economic development and female labor participation have a U-shaped relationship. At lower values of economic development there is a negative relationship with female labor participation, but beyond some level of economic development the relationship becomes positive. Fertility rate and female labor participation have a negative (and linear) relationship.

Units of analysis: 27 countries (aggregate-level data).

Measurements of independent and dependent variables:

Y: Percentage of females who are in the labor force, calculated as number of females in the labor force divided by the total female population over age 15, times 100 (female labor participation).

X_1: GDP per capita in \$1,000s, U.S. current \$ (GDP per capita).

X_2: Number of births per adult female (fertility rate).

Data:

Country ID	Countries	Female Labor Participation (Y)	GDP per Capita in \$1,000s (X_1)	Fertility Rate (X_2)
1	Austria	54.95	44.73	1.49
2	Belgium	47.77	41.26	1.70
3	Burkina Faso	58.37	0.58	5.35
4	Cameroon	71.16	1.39	4.71
5	Canada	60.81	42.35	1.60
6	Denmark	59.31	53.58	1.71
7	Ethiopia	77.00	0.71	4.20
8	Greece	45.70	17.88	1.33
9	Haiti	63.51	0.73	2.92
10	Hungary	47.96	12.82	1.45
11	Iceland	73.17	60.53	1.80
12	Italy	39.63	30.67	1.35
13	Kenya	62.51	1.46	3.85
14	Madagascar	83.41	0.40	4.18

Country ID	Countries	Female Labor Participation (Y)	GDP per Capita in $1,000s ($X_1$)	Fertility Rate (X_2)
15	Nepal	82.77	0.73	2.12
16	New Zealand	63.92	40.33	1.87
17	Norway	60.88	70.89	1.72
18	Poland	49.02	12.43	1.32
19	Puerto Rico	33.25	30.52	1.30
20	Romania	44.48	9.57	1.58
21	Spain	52.30	26.62	1.33
22	Sweden	60.67	51.84	1.85
23	Togo	76.07	0.59	4.45
24	United Kingdom	56.77	40.41	1.80
25	United States	55.99	57.59	1.80
26	Vietnam	73.20	2.17	1.95
27	Zimbabwe	78.21	1.27	3.76

To incorporate the possible curvilinear relationship between GDP per capita and female labor participation in the regression analysis, the researcher needs to create a variable for squared GDP per capita (X_1^2) in her software. The data should look like this:

Country ID	Country	GDP per Capita in $1,000s ($X_1$)	Squared GDP per Capita (X_1^2)
1	Austria	44.73	2,000.77
2	Belgium	41.26	1,702.39
3	Burkina Faso	0.58	0.34
4	Cameroon	1.39	1.93
5	Canada	42.35	1,793.52
6	Denmark	53.58	2,870.82
7	Ethiopia	0.71	0.50
8	Greece	17.88	319.69
9	Haiti	0.73	0.53
10	Hungary	12.82	164.35

(Continued)

Country ID	Country	GDP per Capita in \$1,000s (X_1)	Squared GDP per Capita (X_1^2)
11	Iceland	60.53	3,663.88
12	Italy	30.67	940.65
13	Kenya	1.46	2.13
14	Madagascar	0.40	0.16
15	Nepal	0.73	0.53
16	New Zealand	40.33	1,626.51
17	Norway	70.89	5,025.39
18	Poland	12.43	154.50
19	Puerto Rico	30.52	931.47
20	Romania	9.57	91.58
21	Spain	26.62	708.62
22	Sweden	51.84	2,687.39
23	Togo	0.59	0.35
24	United Kingdom	40.41	1,632.97
25	United States	57.59	3,316.61
26	Vietnam	2.17	4.71
27	Zimbabwe	1.27	1.61

The regression model then includes GDP per capita (X_1) and squared GDP per capita (X_1^2) along with fertility rate (X_2). Plugging in the estimates from the statistical software output into $\hat{Y} = a + b_1 X_1 + b_2 X_1^2 + b_3 X_2$, the estimated regression equation is $\hat{Y} = 60.248 - 1.005 X_1 + 0.016 X_1^2 + 2.987 X_2$. Complete output is as follows:

$$R^2 = 0.522$$
$$SSE = 2,175.637$$

	b	SE of b	t	p-value
Intercept (constant)	60.248	9.208	6.543	<0.001
GDP per capita (X_1)	-1.005	0.398	-2.523	0.019
Squared GDP per capita (X_1^2)	0.016	0.006	2.725	0.012
Fertility rate (X_2)	2.987	2.385	1.253	0.223

R^2 is moderately high (0.522), indicating that 52.2% of the variance in female labor participation (Y) is explained by the set of X's in this model. As for collinearity, the correlations are moderate between fertility rate (X_2) and GDP per capita (X_1) (r = –0.614), and also between fertility rate (X_2) and squared GDP per capita (X_1^2) (r = –0.451). As expected, GDP per capita (X_1) and squared GDP per capita (X_1^2) are very highly correlated (r = 0.957). However, as p-values from t-statistics indicate that the regression coefficients for both GDP per capita (X_1) and squared GDP per capita (X_1^2) significantly differ from zero, and the correlation between X_1 and X_1^2 here is not extremely close to one, centering does not seem necessary in this case. For completeness, the appendix shows the results that are produced when centering X_1 in this example.

It is sensible to consider the highly correlated X_1 and X_1^2 as a set and first test the null hypothesis that both of the population effects for GDP per capita (β_1) and for the squared GDP per capita (β_2) are zero, or H_0: $\beta_1 = \beta_2 = 0$. The researcher can test this hypothesis via an F-test, comparing this regression to another that does not include GDP per capita (X_1) and squared GDP per capita (X_1^2). As before, F is calculated as

$$F = \frac{(SSE_0 - SSE_1)/s}{SSE_1/df_1}$$

SSE_0 = sum of squared errors from the "partial" model that excludes the pair X_1 and X_1^2, here GDP per capita (X_1) and squared GDP per capita (X_1^2), leaving only fertility rate (X_2). Output for this regression is not shown here, but it gave SSE_0 = 2,893.385. SSE_1 is the sum of squared errors from the "full" model that includes all X's, in this case GDP per capita (X_1), squared GDP per capita (X_1^2), and fertility rate (X_2); the output here shows that SSE_1 = 2,175.637. Finally, s is the difference in number of X's between the partial and full models, here two, and df_1 = N – (# of X's in the full model) – 1 = 27 – 3 – 1 = 23.

$$F = \frac{(2,893.385 - 2,175.637)/2}{2,175.637/23} = 3.794$$

With s = 2 and df_1 = 23, the F-table for (2, 23) df (page 237) shows that the p-value for F = 3.794 lies between 0.025 and 0.050. (An online F-calculator gives a more precise p-value of 0.038.) Because this p-value is less than the 0.05 cutoff, the researcher rejects the null hypothesis (H_0: $\beta_1 = \beta_2 = 0$) and concludes that the regression with the X_1 and X_1^2 variables fits better than the regression without them. GDP per capita does appear to have a statistically significant relationship with female labor participation. Further, the specific significance test of $\beta_2 = 0$, for the coefficient on X_1^2, also gave a small p-value (from the t-statistic, with p = 0.012 in the output above). The relationship thus appears to be nonlinear, and the researcher went ahead with interpreting it as such.

The p-value from the t-statistic for fertility rate (X_2) (p = 0.223) indicates that fertility rate (X_2) did not have a statistically significant effect on female labor participation (Y), controlling for GDP per capita (X_1) and squared GDP per capita (X_1^2), and the researcher did not interpret further. Note that the estimate b_3 was in the unexpected positive direction, but again not statistically significant.

The b's for X_1 and X_1^2 have different signs: negative for GDP per capita (X_1) and positive for squared GDP per capita (X_1^2). Because GDP per capita is inherently positive, both signs should be considered in describing the relationship, and they suggest that GDP per capita (X_1) and predicted female labor participation ($\hat{Y}$) do have the expected U-shaped relationship while holding fertility rate (X_2) constant. At lower values of GDP per capita (X_1), there is a negative relationship with the predicted female labor participation ($\hat{Y}$), but beyond some value of GDP per capita (X_1) the relationship becomes positive.

It is important to calculate this value of GDP per capita (X_1) at which the relationship changes sign and check if the value is realistic in light of descriptive statistics for GDP per capita (X_1) in this sample. The estimated value of GDP per capita (in $1,000s) at which the relationship changes sign is calculated as $-b_1/2b_2 = -(-1.005)/(2 \times 0.016) = 31.406$, or 31,406. From the descriptive statistics, the values of GDP per capita (X_1) in the sample vary between $400 (Madagascar) and $70,890 (Norway), with a mean of $24,224. From these figures, $31,406 does seem to be a realistic value and in fact is quite close to the observed GDP per capita for Italy and Puerto Rico ($30,670 and $30,520, respectively).

When GDP per capita (X_1) values are < $31,406, estimates indicate a negative relationship between GDP per capita (X_1) and predicted female labor participation ($\hat{Y}$). In that range, higher values of GDP per capita are associated with lower predicted values of female labor participation. However, as GDP per capita (X_1) increases further, the positive coefficient on squared GDP per capita (X_1^2) becomes most important. For GDP per capita values > $31,406, higher values of GDP per capita are associated with higher predicted values of female labor participation. Increasing wealth seems to go along with decreasing female labor participation for less economically developed countries. On the other hand, increasing wealth is associated with greater female labor participation among economically more developed countries.

It is helpful to illustrate this further by plugging some GDP per capita values (in $1,000s) into the estimated regression equation and looking at the predicted female labor participation ($\hat{Y}$) for each. For the purpose of examining the change in predicted female labor participation ($\hat{Y}$) while controlling for fertility rate (X_2), it is fine to focus only on that part of the regression equation that involves X_1 and X_1^2 (leaving aside the part with a and X_2), as shown earlier. The results are summarized in the following table. Note that none of the fully calculated predicted values for female labor force participation ($\hat{Y}$) would be negative; this is only for the part involving X_1.

GDP per Capita in $1,000s ($X_1$)	Squared GDP per Capita in $1,000s ($X_1^2$)	$-1.005X_1 + 0.016X_1^2$	Change From Previous Predicted Female Labor Participation ($\hat{Y}$)
0	0	0	–
5	25	−4.63	−4.63
10	100	−8.45	−3.82
15	225	−11.48	−3.03

GDP per Capita in \$1,000s (X_1)	Squared GDP per Capita in \$1,000s (X_1^2)	$-1.005X_1 + 0.016X_1^2$	Change From Previous Predicted Female Labor Participation ($\hat{Y}$)
20	400	-13.70	-2.22
25	625	-15.13	-1.43
30	900	-15.75	-0.62
35	1,225	-15.58	0.17
40	1,600	-14.60	0.98
45	2,025	-12.83	1.77
50	2,500	-10.25	2.58
55	3,025	-6.88	3.37
60	3,600	-2.70	4.18
65	4,225	2.28	4.98
70	4,900	8.05	5.77

The calculation above indicated 31.406 as the value of GDP per capita in \$1,000s (X_1) where the relationship changes from negative to positive. In the fourth column of the table, we can see that when GDP per capita in \$1,000s (X_1) is smaller than this, increases in GDP per capita (X_1) lead to decreases in the predicted female labor participation ($\hat{Y}$), with negative values of the change in the predicted female labor participation ($\hat{Y}$) (assuming that the fertility rate (X_2) remains constant). These decreases in predicted female labor participation ($\hat{Y}$) get smaller and smaller as the values of GDP per capita (X_1) get larger. That is, the decrease in the predicted female labor participation ($\hat{Y}$) when GDP per capita in \$1,000s (X_1) increases from 5 to 10 is greater than the decrease in $\hat{Y}$ when X_1 increases from 10 to 15 (a decrease of 3.82 vs. a decrease of 3.03, from the table).

When GDP per capita in \$1,000s (X_1) is 31.406, the relationship changes sign to positive. At this level of GDP per capita (X_1), the value of this part of the equation for predicted female labor participation ($\hat{Y}$) is at its lowest (–15.782: $-1.005(31.406) + 0.016(31.406^2) = -15.782$); because the table has no entry for $X_1 = 31.406$, this is not directly shown, but it is almost like the entry for $X_1 = 30$. Beyond this level, increasing GDP per capita (X_1) leads to an increase in the predicted female labor participation, indicated in the table by positive changes in the fourth column. These increases in predicted female labor participation ($\hat{Y}$) pick up speed as GDP per capita (X_1) continues to increase: the increase in predicted female labor participation as GDP per capita increases from 35 to 40 (0.98) is smaller than the increase in $\hat{Y}$ as X_1 changes from 40 to 45 (1.77), and so on.

The researcher also used this information to assess the real-world significance of GDP per capita's (X_1) effect on predicted female labor force participation ($\hat{Y}$). With GDP per capita (X_1) ranging from 0.40 to 70.89 (in $1,000s) in the sample, the table essentially reflects the whole range of this variable. Differences of 5 or 10 in GDP per capita seem reasonable for assessing the size of that variable's effect. Of course the resulting change in predicted female labor participation depends on where we focus in the table, but many of the changes in $\hat{Y}$ shown in the fourth column of the table seem large relative to the mean, standard deviation, and range of female labor participation in the sample. (These are 60.47, 13.23, and between 33.25 and 83.41, respectively). The estimated effect of GDP per capita (X_1) on female labor force participation (Y) therefore seems meaningful in real-life terms. (As in other examples throughout the text, "effect" is used as a convenient term rather than necessarily implying an actual causal effect.)

6.4 Exercises

6.4.1 Exercise 1: Interpreting With X and X^2

To examine a possibly nonlinear relationship between X_1 and Y, you included X_1 and X_1^2 as independent variables along with a numerical X_2 in your regression. You obtained the following results for that variable, along with sum of squared errors 1,042.56. Your sample size is 34, and in another regression including only X_2, the sum of squared errors was 1,392.45.

	b	p-value
X_1	0.450	0.040
X_1^2	-0.032	0.030
X_2	0.231	0.021

Interpret these results for X_1 and X_1^2 as fully as possible, following what you learned in this chapter.

6.4.2 Exercise 2: Age and Career-Related Networking

For adults, age and time spent on career-related networking have a potentially complicated relationship that might be better characterized by a curve than a straight line. At relatively young ages and during early career stages, we may expect that the relationship between age and time spent on career-related networking is positive. People who are building their careers may increasingly recognize that, in some fields at least, advancement depends in part on becoming known to others working in their area. Further, the importance of such networking may continue as workers gain more experience in their field, so that the incentive to engage in networking remains as workers age and are advancing in their careers. This suggests that at younger ages, age is positively related to time spent in career networking.

However, as workers get older and approach the end of their careers, they may perceive less opportunity for advancement or otherwise reduce their engagement with their field. If so, at older ages there may be a negative relationship between age and time spent in career networking. Together, these arguments suggest that the relationship between age and career-related networking may have an upside-down U-shape. On the other hand, education likely has a simpler straight-line relationship with career-related networking. More educated people tend to spend more time on such activities than the less educated, as they are more likely to be in occupations for which such contact is important in career advancement.

These ideas can be investigated via a multiple regression in which time spent on career-related networking is the dependent variable. For this analysis, career-related networking includes any activities sponsored by professional associations or occupation-related organizations, continuing professional education, and also social events that are specifically aimed at workers in particular fields or regions. Time spent in career-related networking was measured as the total hours spent in such activities per year. Age was measured in years, and education was indicated as years of completed formal education. The following (hypothetical) data were collected for a sample of 22 adults.

Data:

Person ID	Hours Spent on Career-Related Networking	Age	Education
1	78	45	16
2	91	50	13
3	34	20	12
4	23	23	13
5	45	65	11
6	30	70	16
7	80	34	16
8	90	44	20
9	34	27	16
10	50	28	15
11	30	67	17
12	56	25	14
13	100	56	20
14	70	60	18
15	40	28	12
16	58	36	16

(Continued)

Person ID	Hours Spent on Career-Related Networking	Age	Education
17	56	55	12
18	54	60	16
19	100	65	18
20	80	59	16
21	45	32	12
22	40	45	13

1. Analyze the data using multiple regression in a way that incorporates the hypothesized curvilinear relationship between age and time spent on career-related networking.
2. Fully interpret the results of the analysis in Question 1, including, if appropriate, calculation of the estimated age at which the direction of the age-networking relationship changes.

6.5 Appendix

In the example presented in Section 6.3, we mentioned the possibility of centering GDP per capita (X_1) for this analysis. Then the regression would work with GDP per capita in the form of deviations from its sample mean (that is, subtracting the mean 24.224 from each country's GDP per capita) rather than its raw values. With this centering, and with the squared variable X_1^2 calculated from the centered version, regression results are as follows:

$R^2 = 0.522$
SSE = 2,175.637

	b	SE of b	t	p-value
Intercept (constant)	45.219	4.974	9.092	<0.001
Centered GDP per capita (X_1)	-0.236	0.146	-1.611	0.121
Squared centered GDP per capita (X_1^2)	0.016	0.006	2.725	0.012
Fertility rate (X_2)	2.987	2.385	1.253	0.223

We first notice the many results that are the same as in the earlier version of the analysis that used uncentered GDP per capita, including R^2 and the sum of squared errors, and estimates for fertility rate (X_2) and the squared GDP per capita term (X_1^2). These similarities convince us that the analysis really is fundamentally the same as before, but we do note some differences in the output. The intercept is different, though we are not interpreting that directly anyway. More importantly, b_1 and the standard error of b_1 (and in turn the t-statistic and p-value) are different for GDP per capita (X_1) in this centered version. The standard error of b_1 is now smaller (0.146, compared to 0.398 in the uncentered version), as the collinearity introduced by the extremely high correlation between X_1 and X_1^2 is no longer a factor. (In this centered version, the correlation between these X_1 and X_1^2 has been reduced to 0.470, after being 0.957 in the uncentered version.) However, with the estimate b_1 also becoming smaller, the p-value for X_1 is now > 0.05.

Earlier we noted that coefficient estimates that differed between the two versions were related in a systematic way. The previous b_1—here denoted b_1^{un}, for uncentered—is related to b_1 in the centered version by the expression $b_1^{un} = b_1^{cent} - 2\overline{X}_1 b_2$. To see this exactly we need more decimal places than the output displays here, particularly for b_2, but we can see this approximately as $-0.236 - (2 \times 24.224 \times 0.016) = -1.011$, which is close to −1.005. Similarly, the intercepts are related by $a^{un} = a^{cent} - b_1^{cent}\overline{X}_1 + b_2(\overline{X}_1)^2$. Again we would need more decimal places to see this exactly, but $45.219 - (-0.236 \times 24.224) + (0.016 \times 24.224^2) = 60.325$, not far from 60.248. This reinforces the idea that these two analyses are not fundamentally different. Also, as the sum of squared errors is the same for the two analyses, the F-test for the pair of coefficients will be the same as well. Both analyses suggest that the addition of the X_1 and X_1^2 terms improves the model. Combined with the statistical significance of the X_1^2 term, there does seem to be a U-shaped relationship between GDP per capita (X_1) and female labor participation (Y).

Finally, these results also indicate the same value of GDP per capita (X_1) at which the relationship changes sign, except that it is expressed as a deviation from the mean: $0.236 / (2 \times .016) = 7.375$. With the mean of GDP per capita (X_1) equal to 24.224, this deviation is equivalent to 31.599 in the original values. The small difference from 31.406 again stems from rounding the output.

7

Categorical Dependent Variables

Logistic Regression

In Chapter 3, we discussed categorical independent variables and the construction of dummy variables. It is easy to see that there are also many research questions for which it would be important to consider a categorical dependent variable. For example, a research study might investigate the factors that predict a voter's identification with one political party over another or the characteristics that predict whether or not an individual is unemployed. In both of these cases, the dependent variable would be categorical rather than numerical in nature (Republican or Democratic; employed or unemployed). As with categorical independent variables, we need special strategies to deal with this kind of nonnumerical dependent variable. But unlike the situation for categorical independent variables, we cannot address this simply by creating dummy variables. This chapter will explore *logistic regression*, one of the most common methods for regression analysis involving a dichotomous (two-category) dependent variable. As before, we provide a full example and exercises at the end of the chapter. Also, we give a brief introduction to some extensions of logistic regression such as multinomial and ordinal (ordered) logistic regression and the alternative of probit regression.

7.1 Concerns With Multiple Regression for a Dichotomous Dependent Variable

When our research goals require us to think about a dichotomous dependent variable, our first instinct might be to simply carry what we learned about dummy variables for categorical independent variables over to this new setting

FIGURE 7.1 ● A Simple Regression With a Zero-One Dependent Variable

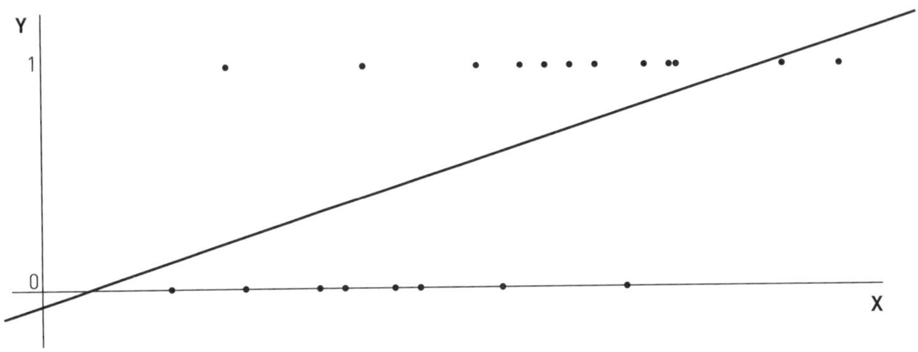

of a dichotomous dependent variable. That is, if employment status (measured as employed or unemployed) were our dependent variable, we could simply create the usual sort of 0–1 dummy variable representing employment status that we learned before and then use this as the dependent variable in our multiple regression.

However, it seems unsatisfactory to use a straight-line regression model in the case of a 0–1 dependent variable. Figure 7.1 gives an example of a simple regression involving a 0–1 dependent variable and a continuous numerical independent variable X. Data like these do not have the familiar "cloud of points" appearance that is typical when both X and Y are continuous variables, as we saw in the plots in Chapter 1. The notion of fitting a straight line to points in a plot like this seems a bit strange, as the pattern of points does not bring to mind a straight line. It does seem sensible that the regression line has a positive slope, as it looks like higher values of Y (that is, Y = 1) become more prevalent for larger values of X, but the line does not represent the general pattern of points very well. The only possible values that Y can take in the data are 0 and 1, but the predictions implied by the straight line are not so limited. There are values of X for which the predicted Y is negative, or > 1, or any of the possible decimal values between 0 and 1. The restriction that the values of Y can be 0 or 1 only is not acknowledged in any way by this regression, which seems undesirable for an analysis in which one of the primary goals is to predict Y.

There is no easy solution to this problem if we stay within the usual regression framework of predicting Y. However, it is possible to shift our focus. Rather than the analysis aiming to predict the value of Y directly, we can instead focus on predicting the *probability* that Y is equal to 1. That is, we look at the probability that a case's value of Y falls into the category labeled one rather than the category labeled zero, instead of predicting Y's numerical value of 0 or 1. Because probability by definition ranges between 0 and 1, legitimate possible values for P{Y = 1} include all the possible decimals between 0 and 1. This would address one of our concerns with analyzing Y directly.

We might next simply reinterpret a plot like that in Figure 7.1 as predicting P{Y = 1} rather than the value of Y directly. Such a regression, sometimes including refinements in how the estimated b's are obtained, is then called a *linear probability model*, with the b's in this approach interpreted as showing effects on the probability that a case has Y = 1. This linear probability model can give quite reasonable results in a variety of practical situations, and this approach still has its fans today. However, in many social science fields it is not a favored approach to analysis of a dichotomous dependent variable. One reason is that this model still allows the possibility of predicted values lying outside the allowed 0–1 range of a legitimate probability. In the plot in Figure 7.1, low values of X give a predicted probability < 0, and high values give a predicted probability > 1.

Rather than reinterpreting the usual regression this way, an alternative is to build a model that is right from the start explicitly directed at P{Y = 1}. We focus on one popular way of doing this, but there are also other alternatives that we do not explore. This shift to focusing on P{Y = 1} requires a somewhat different framework than we have examined with multiple regression in the preceding chapters. We will not discuss many mathematical details of how this framework differs, but more advanced texts can be consulted for those.

We first note the restriction that P{Y = 1} cannot legitimately exceed 1. This is hard to overcome in a straight-line model for P{Y = 1}, leading to consideration of *odds* rather than the probability itself. In everyday language, "odds" may be a synonym for "probability," in remarks like "What are the odds that it will rain today?" But the formal definition of odds involves a comparison of P{Y = 1} and 1 – P{Y = 1}; for a dichotomous Y, the latter is of course P{Y = 0}, but the definition is a bit more general if we use 1 – P{Y = 1}. Then the odds of Y = 1 are defined as the ratio of those two probabilities: P{Y = 1}/(1 – P{Y = 1}). Many people encounter this more precise usage of "odds" in the context of gambling, though there the typical statement refers to the odds *against* some event occurring, rather than the odds *of* that event occurring.

When P{Y = 1} is small (close to 0), (1 – P{Y = 1}) will be close to 1, and this ratio—the odds—will also be small. When P{Y = 1} is large, meaning close to 1, (1 – P{Y = 1}) will be close to 0, and the odds will also be large. For example, if P{Y = 1} = 0.07, then (1 – P{Y = 1}) = 0.93, and the odds would be 0.07/0.93 = 0.075. But if P{Y = 1} = 0.93, and (1 – P{Y = 1}) = 0.07, the odds would be 0.93/0.07 = 13.286. The closer that P{Y = 1} gets to 0, the smaller the odds, but the odds never go below 0. The closer that P{Y = 1} gets to 1, the larger that the odds get, so that they will be extremely large when P{Y = 1} is very close to 1, with no limit on how large they can be: if P{Y = 1} were equal to 0.999, the odds would be 0.999/0.001 = 999. So if our regression model is predicting odds, there is no concern about impossibly high predictions. Any positive number is allowed for the odds and can be converted back to a legitimate probability in the 0–1 range.

This shift to focusing on odds in our regression model, however, still leaves a concern over negative predicted values, as odds cannot legitimately be negative. In response, the focus can shift again, to the *log odds* or *logit*; applications use the base e (natural log) when logging in this context. For log odds, any positive or negative number is permitted. When the log odds are negative, the implied odds are < 1, and the implied probability is < 0.50. When the log odds are positive, the implied odds are > 1, and the implied probability is > 0.50. If the log odds are 0, then the odds are equal to 1 (remember that log 1 = 0), and the probability is equal to 0.50. A model for the log odds, then, does not have to worry about impossible predictions.

7.2 Logistic Regression

This sort of thinking leads to the *logistic regression* model for dichotomous dependent variables, often referred to as the logit model for short. It is defined as a linear model for the log odds (logit) that Y equals 1; note that our choice of which category of Y is labeled 1 and which is labeled 0 is arbitrary and does not fundamentally affect our results. Writing the "true" logit regression model using the true effects—that is, the β's in the population that our software will estimate as b's from the sample data—with three independent variables, the model is

$$\log\frac{P\{Y = 1\}}{1 - P\{Y = 1\}} = \alpha + \beta_1 X_1 + \beta_2 X_2 + \beta_3 X_3.$$

The model holds that each case in our data has some underlying $P\{Y = 1\}$, and that when this probability is converted to log odds, it is related to the independent variables via the expression $\alpha + \beta_1 X_1 + \beta_2 X_2 + \beta_3 X_3$. This expression looks familiar from our previous discussions of multiple regression for a continuous Y, where it implies a straight-line relationship between an X and the dependent variable, holding other X's constant. That same idea of a linear model applies here, but the role of dependent variable is being played by the log odds that Y is equal to 1, rather than the predicted value of a continuous Y as before. Although it is still natural to think of Y as our dependent variable when we are collecting data, we need to keep in mind that the formal model does not directly relate the independent variables to a case's observed value of Y, but rather to the case's log odds that Y equals 1.

The model can also be expressed directly in terms of $P\{Y = 1\}$, because the log odds are a function of that probability. The notation will be easier to follow if a single letter (here we use L, for logit) represents the right-hand side of the logit model expression above, so that $L = \alpha + \beta_1 X_1 + \beta_2 X_2 + \beta_3 X_3$, and the logit model is now expressed as $\log\frac{P\{Y = 1\}}{1 - P\{Y = 1\}} = L$. The appendix works through the algebra steps showing how this implies that

$$P\{Y = 1\} = \frac{e^L}{1 + e^L}$$

A simple example can help illustrate how the "straight-line" linear model for the log odds of $\{Y = 1\}$ implies a nonlinear model for the probability of $\{Y = 1\}$. Suppose that the true model for the log odds is given by $\log\frac{P\{Y = 1\}}{1 - P\{Y = 1\}} = -1.36 + 0.22\,X$, and further suppose that the X variable is such that its realistic values range between 0 and 20. The graph of this relationship between the logit (log odds, labeled L in the graph) for $\{Y = 1\}$ and X, over the range of realistic X values, is shown in Figure 7.2. As expected, the graph shows a straight line.

We can plot the same relationship with X, but putting the implied probabilities $P\{Y = 1\} = \frac{e^L}{1 + e^L}$, instead of the logit, on the vertical axis in Figure 7.3. This is clearly not a straight line, so there is not a linear relationship between X and $P\{Y = 1\}$. For the particular range of X values roughly between 4 and 9, the curve does look rather straight, but as the X values get larger, a given increase in X produces less and less of an increase in $P\{Y = 1\}$. This makes sense in light of

FIGURE 7.2 • Plot of $\log\frac{P\{Y = 1\}}{1 - P\{Y = 1\}} = -1.36 + 0.22\,X$

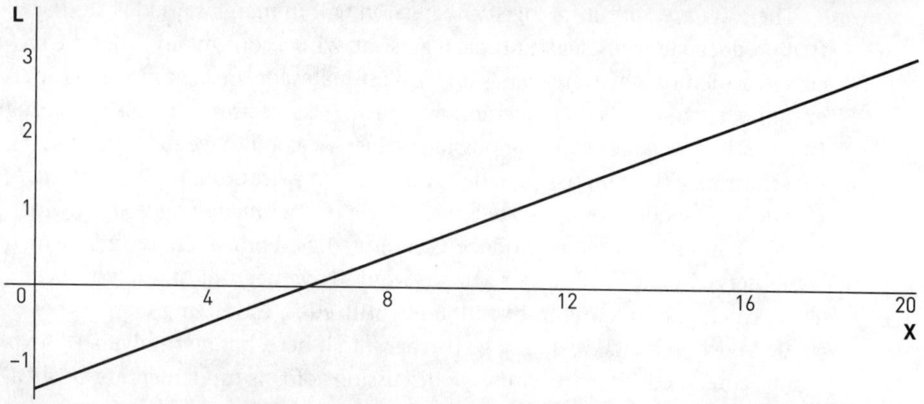

FIGURE 7.3 • Plot of P{Y = 1} as Calculated From $P\{Y = 1\} = \frac{e^L}{1 + e^L}$

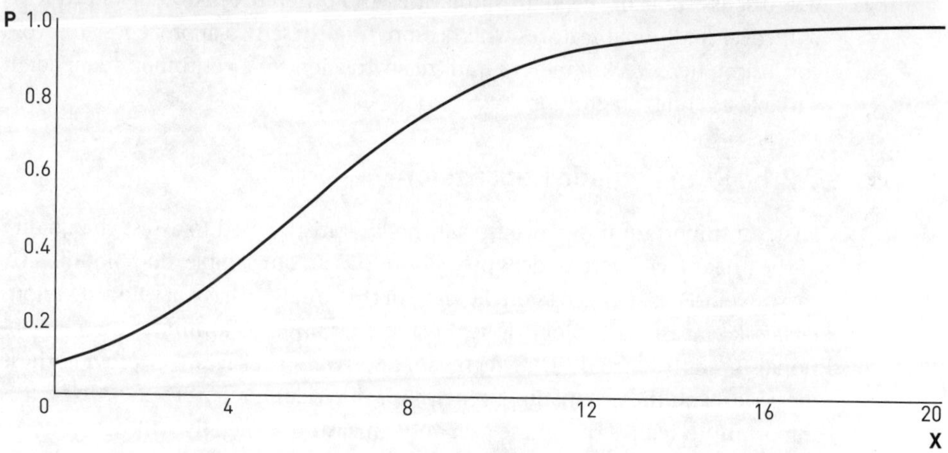

the restriction that P{Y = 1} cannot exceed one; it would be impossible for P{Y = 1} to keep increasing as it does when X is between 4 and 9 without violating this restriction. So the linear model for the logit translates into a nonlinear model for P{Y = 1}.

Our statistical software will report estimates a and b_1, b_2, and b_3 for the unknown α and β's in the logit model at the beginning of this section. (Of course, we can certainly have more than three independent variables in our analysis.) To be consistent with earlier notation, we could put a hat on $\log\frac{P\{Y = 1\}}{1 - P\{Y = 1\}}$ when discussing the estimated model that uses a, b_1, b_2, b_3, . . . instead of α and the β's, like $\widehat{\log\frac{P\{Y = 1\}}{1 - P\{Y = 1\}}}$, or perhaps instead with smaller hats on P{Y = 1} in the expression. Because that becomes rather inconvenient, we will not use the hat in the remainder of the chapter, but still keep in mind that when the model is presented in terms of a, b_1, b_2, b_3, and so on, we are working with estimated logits.

The software output for logistic regression will in many ways look similar to that produced for the usual multiple regression with a continuous Y. But the process by which the software comes up with estimates for the logit model is different. In particular, it is no longer following the least squares principle of finding those estimates, among all the possible values we could have for a and the b's, that minimize the sum of squared residuals in our regression. Instead, estimates for the logit model are motivated by the idea of *maximum likelihood,* in which the estimates are chosen to produce a set of predicted probabilities under which the observed data for Y in our sample would have been most likely to occur. This principle of maximum likelihood is utilized in estimating many types of models beyond logit. We do not go further into it here, but more advanced texts can be consulted for more extensive discussions of this fundamental statistical concept. One practical implication is that, depending on the mix of zero and one values for Y in the sample, larger sample sizes are in general relatively even more desirable than with a continuous Y and least squares.

Some aspects of interpretation for the logistic regression model are like those in multiple regression of a continuous dependent variable, while some differ. We will discuss some of the main similarities and differences now; advanced texts will cover additional features when comparing these two approaches. To avoid confusion, here we will refer to multiple regression for a continuous dependent variable as "linear regression."

7.2.1 R^2 in Logistic Regression

First, an important difference is that the R^2 statistic, used to assess the quality of the linear regression model's predictions of Y in our sample, does not directly carry over to logistic regression. Because of this, analysts have developed various *pseudo-R^2* measures for logistic regression that attempt to mimic the conventional R^2 statistic from linear regression. Software packages will provide these measures, and these statistics often appear in research reports and articles. It is important to realize that these pseudo-R^2 measures generally will not share all the familiar properties of the linear regression R^2. Interpretation is accordingly not always straightforward, and researchers vary in how much trust they place in these measures. The pseudo-R^2 measures, therefore, do not take the central role in logistic regression that the regular R^2 statistic plays in interpretation of linear regression results.

As an alternative assessment of the model's performance, software output will often include a 2×2 *classification table,* in which the cases in the sample are grouped according to (a) whether a case's estimated $P\{Y = 1\}$ from the model was above or below 0.50 and (b) whether the actual Y for that case was 0 or 1. The idea is that the model in a sense "got it right" if it estimated $P\{Y = 1\} > 0.50$ for a case that actually had Y = 1, or estimated $P\{Y = 1\} < 0.50$ for a case that actually had Y = 0. On the other hand, it seems that the model "got it wrong" if it estimated $P\{Y = 1\} < 0.50$ for a case that actually had Y = 1, or estimated P{Y =

1} > 0.50 for a case that actually had Y = 0. The classification table summarizes this comparison for all the cases in the sample, showing the proportion of the sample cases in which the estimated probability and the observed Y agreed. A higher proportion suggests better performance by the logit model.

This is a reasonable approach, but one criticism is that if a case has an estimated P{Y = 1} that is very close to 0.50, then the comparison of this estimated probability to the case's observed Y is probably not very informative. Such instances may be better left out when assessing model performance. Another concern is that it is easier for the model to "look good" when values of 0 and 1 for Y in the sample are very imbalanced (i.e., a high proportion of 0's or a high proportion of 1's), but that baseline is not obvious when interpreting the usual table in the software output. So, again, for the logit model it is difficult to come up with a completely satisfactory stand-in for the R^2 from linear regression.

7.2.2 Statistical Significance in Logistic Regression

Next, the assessment of statistical significance of a single independent variable's "effect" (again not necessarily implying a genuine causal effect) is essentially the same as it was in linear regression. The standard error is still meant to represent sampling variability in b, and the familiar ratio of b divided by its standard error will still be used as a test statistic. A minor difference is that this will usually be referred to as a z- rather than t-statistic, with the p-value obtained from a z- rather than t-table, but software output will provide a p-value for H_0: $\beta = 0$ vs. H_1: $\beta \neq 0$ as usual. If required, we can again adapt the test for a null hypothesis involving some other value of β. So this aspect of the application of logistic regression feels very familiar.

However, comparison of nested models, or testing a hypothesis that multiple independent variables have zero effects, will be carried out somewhat differently than in linear regression. Recall that we used such tests to assess statistical significance of a categorical independent variable that required more than one dummy. We also noted some other applications such as tests involving interaction and squared independent variables. These purposes remain important in the logistic regression setting, but the mechanics of the hypothesis test changes, so that we will not be calculating an F-statistic like before. Instead, we will calculate a *chi-square statistic*, and obtain the p-value by looking up the calculated chi-square statistic in a chi-square table (page 247) or using an online calculator or a similar function in statistical software. As in the F-statistic, calculation will focus on the difference in a particular statistic between the two nested models. But the sums of squared prediction errors that were the basis of the F-statistic will not be used in this setting.

Calculation of this chi-square statistic will instead require the *log-likelihood* provided in the software output for each model. We noted previously that the estimated b's in a logit model are obtained under the principle of maximum likelihood, and the number reported for the log-likelihood on the output reflects this maximization. Glossing over many technical issues, the likelihood essentially represents the probability of obtaining the observed sample data on Y if we were to assume that the estimated b's are in fact the true values of these coefficients, with the probabilities of Y = 1 that are implied by these coefficients. The log-likelihood is in turn just the log of this likelihood. Although the log-likelihood does not have a standard interpretation on its own, it is the basis of nested model comparison.

For two nested models labeled Model 0 (with fewer independent variables) and Model 1 (with more independent variables), we will refer to their log-likelihoods as log-likelihood$_0$ and log-likelihood$_1$. The chi-square statistic for testing the null hypothesis that the additional β's in Model 1 are all zero—that is, that Model 1 does not improve on Model 0—is calculated as (–2 × log-likelihood$_0$) – (–2 × log-likelihood$_1$). The difference in the number of independent variables between the two models (the number of β's being tested) will be the degrees of freedom for the *chi-square test*.

Note that some software will report a model's log-likelihood directly, and other software will include the multiplication by –2 in the reported figure. Be sure that you are clear on which protocol your software is following before executing the above chi-square calculation. One fact that may help avoid confusion is that the log-likelihood will be at least somewhat greater for the (larger) Model 1 than Model 0, meaning that, due to multiplication by a negative number, (–2 × log-likelihood$_1$) will be smaller than (–2 × log-likelihood$_0$), and the resulting chi-square statistic will be positive. This is reminiscent of the fact that the sum of squared errors in linear regression will be less for the larger of two nested models. In both cases, using more variables allows a closer fit of model to data.

Suppose that we are conducting a logistic regression in which the dependent variable is political party identification (Republican = 1, Democratic = 0), and the independent variables are education (years), age, gender (two categories, expressed as a dummy variable in which female = 1 and male = 0), and religious identification (with five categories). The statistical significance of education, income, and the single gender dummy can each be assessed via a p-value from the z-test as explained above. However, religious identification requires a set of four dummies, and we therefore need to use the chi-square test to test the hypothesis that all four β's associated with this set of dummies are zero. The chi-square statistic is calculated as (–2 × log-likelihood$_0$) – (–2 × log-likelihood$_1$). Model 0 leaves out the set of four religious identification dummies, so it includes only education, income, and the gender dummy (three X's), while Model 1 includes these plus the set of four dummies for religious identification (seven X's).

Suppose that our software reported –3,426.82 as the log-likelihood for Model 0 and –3,420.33 as the log-likelihood for Model 1. (As expected, the figure is larger, meaning less negative, for Model 1.) Then chi-square is calculated as (–2 × log-likelihood$_0$) – (–2 × log-likelihood$_1$) = (–2 × –3,426.82) – (–2 × –3,420.33) = 6,853.64 – 6,840.66 = 12.98. The nested models differ by four independent variables, the four religious identification dummies, so we compare this figure to the chi-square table for 4 df. According to the chi-square table on page 247, our calculated value of 12.98 lies between the table values 11.143, corresponding to a p-value of 0.025, and 13.277, corresponding to a p-value of 0.01. Our chi-square statistic therefore shows that $0.01 < p < 0.025$. (An online calculator gives a more precise p-value of 0.011.)

Because the p-value is less than the usual cutoff of 0.05, we conclude that the religious identification dummies significantly improved the model, so that religious identification is significantly associated with political identification. This situation of testing a set of dummies in logistic regression is, of course, a very common application of this test, but again, as with the F-test in linear regression, it can be also be used for any other comparison of nested models.

7.2.3 Interpretation of b in Logistic Regression

It is tempting to make an inappropriate interpretation of b's in logistic regression, and there are many examples of published research reports that fall into this trap. The main danger lies in analysts' trying to make a simple interpretation that addresses P{Y = 1} rather than the odds or log odds. Unfortunately, the nature of the logit model means that interpretations involving probability are necessarily somewhat complicated. Remember how the model was re-expressed as $P\{Y = 1\} = \frac{e^L}{1 + e^L}$ and that this meant that the model for the probability is nonlinear. In light of this, it is especially important that we be careful to make correct interpretations of b.

To illustrate this interpretation, we will focus on a sample estimate b_1. The logit model is written for the log odds that {Y = 1}, so b_1 shows the estimated change (an increase if b_1 is positive, a decrease if it is negative) in the log odds when X_1 increases by one unit, holding other independent variables constant. If X_1 is a dummy, then b_1 is the estimated difference in log odds between the dummy's named category and the reference category, again holding other independent variables constant. These interpretations seem very straightforward, but they may not be so useful when we disseminate our findings, because readers of our research are unlikely to have very strong intuition about the meaning of changes in the log odds and therefore will find it hard to make sense of our results.

Recall the similar concern with regression models for log Y in Chapter 5. In response, we focused on interpretations of multiplicative effects on Y, rather than on additive effects on log Y. The same sort of shift is helpful here. Using the L notation introduced earlier in this chapter, the logit model $\log\frac{P\{Y = 1\}}{1 - P\{Y = 1\}} = L$ for the log odds is equivalent to the model $\frac{P\{Y = 1\}}{1 - P\{Y = 1\}} = e^L$ for the odds. Though we did not use the L notation in Chapter 5, the right-hand side of this expression is like the expression in Chapter 5 for the predicted Y when using log Y. In Chapter 5, that led us to interpret e^{b_1} as the multiplicative change in the predicted Y when X_1 increases by one unit. The same thing happens here, except that e^{b_1} represents not the multiplicative change in predicted Y but rather the multiplicative change in the odds of {Y = 1} when X_1 increases by one unit, holding other independent variables constant. In more detail, after increasing X_1 by one unit, $L = a + b_1 X_1 + b_2 X_2 + b_3 X_3$ becomes $L^* = a + b_1(X_1 + 1) + b_2 X_2 + b_3 X_3 = a + b_1 X_1 + b_1 + b_2 X_2 + b_3 X_3$, or the previous $L + b_1$. That means that the new estimated odds $e^{L^*} = e^{(L + b_1)} = e^L e^{b_1}$, or the previous odds times e^{b_1}, show the multiplicative effect of a one-unit increase in X_1 on the estimated odds of {Y = 1}.

In the context of the logistic regression model, e^{b_1} is often referred to as the *odds ratio* for X_1. The reason for this language is easiest to see when X_1 is a dummy variable. Then the odds that {Y = 1} in the named category are estimated to be

e^{b_1} times those in the reference category. That is, the ratio of the odds of {Y = 1} between the named and the reference category is e^{b_1}, hence the name. This connection is a little less clear when X_1 is a numerical variable, but there e^{b_1} is the ratio of the odds of {Y = 1} after increasing X_1 by 1 unit ($X_1 + 1$) to the odds before increasing X_1 by 1 unit (X_1).

Continuing our previous example of logistic regression for political identification, suppose that age and gender (that is, the female dummy) showed statistically significant effects on dichotomous political identification (Republican vs. Democratic). Suppose that the b for age was 0.034 and the b for the female dummy was –0.103. Again these b's are effects on the log odds of identifying as Republican (because Y was set to 1 for Republicans), and we would prefer that our interpretations focus on odds rather than log odds. The odds ratio for age is $e^{0.034} = 1.035$, so as age increases by one year, holding other X's constant, the estimated odds of identifying as Republican are multiplied by 1.035; this can also be expressed as a 3.5% increase in the estimated odds. For gender, the odds ratio for the female dummy is $e^{-0.103} = 0.902$. Holding other independent variables constant, the estimated odds of Republican identification for women are 0.902 times those for men (the reference category). In other words, the estimated odds for women are 9.8% lower than those for men. These calculations are just what we saw in Chapter 5 for linear regression with a logged dependent variable, but in the context of logistic regression they now refer to multiplicative effects on odds.

As in linear regression, it is important to consider the question of real-world significance for such results. But instead of assessing the size of a change in the predicted Y, we are now assessing a multiplicative effect on the odds that Y = 1, and we can consider the size of that change without referring to descriptive statistics for Y. However, we still may want to evaluate real-world significance in light of some change in a numerical X other than the usual one-unit increase. As in Chapter 5, we can easily do that by multiplying b by the desired change in X and then raising e to that power. In the example, suppose that we wanted to consider the impact of a five-year increase in age. We then calculate (0.034 × 5) = 0.170, and $e^{0.170} = 1.185$. (Alternatively, this can be obtained by $1.035^5 = 1.188$, with the small difference due to rounding back in the original calculation of $e^{0.034} = 1.035$.) We therefore estimate that a 5-year increase in age multiplies the odds of Republican identification by 1.185, or an 18.5% increase. Such an increase sounds substantial in real terms.

The multiplicative interpretation of the odds ratio is more accessible to readers than the additive interpretation of b for the log odds, so this shift will make our results more comprehensible to the consumers of our research. However, there is still an obstacle to understanding, because most readers will have a much better intuitive grasp of probability than they do of odds. This problem is compounded by the everyday usage of "odds" as synonymous with "probability." Regrettably, some authors of research reports try to sidestep this by simply misinterpreting odds ratios as if they referred to differences in the probability P{Y = 1} rather than in the odds $\frac{P\{Y = 1\}}{1 - P\{Y = 1\}}$. In our example, this mistake would mean interpreting the 9.8% lower *odds* of Republican identification for women as 9.8% lower *probability*. Even if odds and probability might be used as roughly synonymous terms in everyday life, in the more exacting language of research they mean very different things. A further confusion is that although the proper statement about odds is describing a multiplicative change, the improper application to probability sounds more like an additive change (which makes it doubly incorrect).

This misinterpretation is encountered surprisingly often, but it is quite poor research practice that should be avoided. Note that when we are working in a data context for which P{Y = 1} would realistically always be very small, probability and odds do not differ much for probabilities that are very close to zero. In that case, one would be more justified in interchanging probability and odds, but it would still need to be clear that multiplicative rather than additive changes are being reported with the odds ratio.

Legitimate interpretations in terms of probability are possible, but these will require more calculation and also carry the caveat that specific statements about changes in probability are correct for a particular kind of case, or reflect some kind of averaging over cases, rather than being universally correct in each specific case. (This again parallels the discussion in Chapter 5 of interpretation when Y has been logged.) By a "particular kind of case" we mean a particular profile of X values that form a backdrop for examining how changes or differences in one X variable affect the estimated P{Y = 1}.

Continuing with our example, we may be especially interested in how differences in religious identification are related to the probability of identifying as Republican. To do this, we could specify values for all of the other X variables in our logit model, here education, age, and gender. For numerical variables like education and age it is common to use the variables' means, and for categorical variables like gender it is common to use the variable's modal (most frequent) category. Or there may be a particular profile of X values that is especially instructive to consider, perhaps representing a specific type of case that plays an important role in theoretical ideas about our research topic. With these specified values for education, age, and gender, and the estimated b's from our output, we can calculate an estimate of $P\{Y = 1\} = \frac{e^L}{1 + e^L}$ for each religious category. It is easy to see that this kind of illustration of how the estimated probability of Republican identification changes across the religious categories could enhance our understanding of the logistic regression results.

We could also use this approach to see how changes in a numerical variable like age affect the estimated probability of Republican identification, by fixing values for education, gender, and religious identification and calculating an estimated P{Y = 1} for values across age's range in our sample. In either case, we would need to be very clear that these findings are specific to whatever values we chose to use for the other X's, and therefore cannot be taken as universal differences in probability associated with differences in gender or age.

Some software will give us options for other variations on this strategy. For instance, we might consider the actual values of education, age, and religious identification for each person in our sample. We could then calculate for each case the estimated probability of Republican identification for that configuration of values with gender set to female, as well as the estimated probability of Republican identification for that configuration with gender set to male. Averaging the difference in these estimated probabilities over our entire sample would be another way to express the typical gender difference in probability of Republican identification. Again, we would need to make clear how this calculation was made and how it relied on values for other independent variables. There are various other approaches to this problem as well, and software documentation can be consulted to see what is available. We will illustrate some of the simpler probability calculations in the following example.

7.3 Example: Cessation of Smoking

This hypothetical example will show how to carry out and interpret a logistic regression analysis involving a dichotomous (two-category) dependent variable. A research study included 75 adults, age 40 years and older, who were participating in a quit-smoking program for employees at a university. (This N is somewhat larger than in previous chapters' examples; this reflects a desire to have a larger sample size for logistic regression.) The program director planned to examine the effects of past attempts to quit smoking, education, and social support on smoking cessation. She hypothesized that the more times in the past that someone has tried but failed to quit, the less likely it will be for them to succeed in quitting. She also expected that a participant with more education will have a greater chance of successful cessation and that social support will also be related to smoking cessation.

Smoking cessation was indicated by the dichotomous outcome of "successful" and "unsuccessful." Success was defined as not smoking during the program and a 6-month follow-up period. Past attempts to quit smoking were the number of attempts that the subject had ever made during their adult life. Education was measured by years of formal schooling. Social support was indicated by the subject's intimate relationship status during the program and the follow-up period, categorized as "married," "unmarried partner," and "single."

The research hypothesis, units of analysis, measurements of independent and dependent variables, and raw data are as follows:

> *Research hypothesis*: More frequent past attempts are associated with a lower chance of successful cessation. Greater education is associated with greater chance of successful cessation, and people in different relationship statuses will have different chances of success.
>
> *Units of analysis*: 75 adults over 40 years old (individual-level data) who participated in the quit-smoking program.
>
> *Measurements of independent and dependent variables:*
>
> Y: Smoking cessation: "Successful" or "Unsuccessful" (cessation).
>
> X_1: Number of past attempts to quit smoking (attempts).
>
> X_2: Years of formal schooling (education).
>
> X_3: Intimate relationship status: "Married," "Unmarried partner," "Single" (intimate relationship).

Data:

Person ID	Cessation (Y)	Past Attempts (X_1)	Education (X_2)	Intimate Relationship (X_3)
1	Successful	4	22	Unmarried partner
2	Unsuccessful	14	16	Single
3	Successful	2	16	Unmarried partner
4	Unsuccessful	18	9	Unmarried partner
5	Unsuccessful	9	12	Married
6	Unsuccessful	12	12	Single
7	Successful	0	18	Unmarried partner
8	Successful	0	15	Unmarried partner
9	Successful	0	21	Unmarried partner
10	Successful	2	13	Single
11	Unsuccessful	5	16	Single
12	Successful	2	18	Unmarried partner
13	Unsuccessful	8	12	Unmarried partner
14	Successful	8	12	Single
15	Unsuccessful	11	9	Unmarried partner
16	Unsuccessful	0	14	Single
17	Successful	4	24	Married
18	Unsuccessful	6	12	Single
19	Successful	9	22	Unmarried partner
20	Successful	10	24	Married
21	Successful	6	22	Unmarried partner
22	Unsuccessful	2	18	Single
23	Unsuccessful	1	16	Single
24	Unsuccessful	8	14	Unmarried partner
25	Unsuccessful	0	12	Married
26	Successful	14	20	Unmarried partner
27	Successful	3	18	Unmarried partner
28	Successful	5	16	Unmarried partner
29	Successful	12	18	Married
30	Unsuccessful	4	16	Single

(*Continued*)

Person ID	Cessation (Y)	Past Attempts (X_1)	Education (X_2)	Intimate Relationship (X_3)
31	Successful	6	14	Married
32	Successful	3	9	Unmarried partner
33	Unsuccessful	0	18	Unmarried partner
34	Successful	4	17	Unmarried partner
35	Unsuccessful	16	22	Unmarried partner
36	Successful	0	14	Married
37	Unsuccessful	10	18	Unmarried partner
38	Unsuccessful	10	20	Unmarried partner
39	Successful	10	15	Married
40	Successful	12	17	Single
41	Unsuccessful	8	11	Unmarried partner
42	Successful	1	20	Single
43	Successful	4	22	Married
44	Successful	5	18	Married
45	Unsuccessful	8	16	Single
46	Successful	2	24	Unmarried partner
47	Unsuccessful	10	16	Married
48	Unsuccessful	2	18	Single
49	Unsuccessful	1	22	Single
50	Successful	3	10	Unmarried partner
51	Successful	0	20	Married
52	Successful	2	24	Married
53	Successful	3	22	Unmarried partner
54	Successful	4	20	Unmarried partner
55	Successful	4	21	Married
56	Unsuccessful	2	22	Married
57	Successful	9	12	Single
58	Unsuccessful	2	18	Single
59	Unsuccessful	1	20	Married
60	Unsuccessful	18	9	Married
61	Successful	2	20	Single
62	Unsuccessful	12	10	Single

Person ID	Cessation (Y)	Past Attempts (X_1)	Education (X_2)	Intimate Relationship (X_3)
63	Successful	1	19	Unmarried partner
64	Successful	1	16	Unmarried partner
65	Unsuccessful	8	22	Single
66	Successful	1	24	Married
67	Unsuccessful	14	16	Married
68	Successful	9	18	Single
69	Successful	0	18	Unmarried partner
70	Successful	0	20	Unmarried partner
71	Unsuccessful	4	16	Single
72	Unsuccessful	3	13	Unmarried partner
73	Successful	1	18	Married
74	Unsuccessful	10	17	Unmarried partner
75	Successful	2	20	Married

Because smoking cessation, the dependent variable, is dichotomous (has two categories), the program director decided to use logistic regression. She assigned a value of 0 to unsuccessful cessation outcomes and a value of 1 to successful outcomes. Although most statistical software accepts a binary categorical dependent variable in the logistic regression procedure even if the values are presented as words, recoding to 0 and 1 helps avoid later confusion in interpreting the software output. For the categorical independent variable intimate relationship status (X_3), she selected "Single" as the reference category (as before, the selection of the reference category does not influence the results in a fundamental way) and created dummy variables for "Married" and "Unmarried partner." The married (X_{3a}) and unmarried partner (X_{3b}) dummy variables together represent the categorical variable intimate relationship status (X_3). The recoded cessation variable and the dummy variables for intimate relationship are shown in the following table:

				Dummy Variables for Intimate Relationship	
Person ID	Cessation (Y)	Cessation (Y)	Intimate Relationship (X_3)	"Married" Dummy (X_{3a})	"Partner" Dummy (X_{3b})
1	Successful	1	Unmarried partner	0	1
2	Unsuccessful	0	Single	0	0
3	Successful	1	Unmarried partner	0	1
4	Unsuccessful	0	Unmarried partner	0	1

(Continued)

Person ID	Cessation (Y)	Cessation (Y)	Intimate Relationship (X_3)	Dummy Variables for Intimate Relationship	
				"Married" Dummy (X_{3a})	"Partner" Dummy (X_{3b})
5	Unsuccessful	0	Married	1	0
6	Unsuccessful	0	Single	0	0
7	Successful	1	Unmarried partner	0	1
8	Successful	1	Unmarried partner	0	1
9	Successful	1	Unmarried partner	0	1
10	Successful	1	Single	0	0
11	Unsuccessful	0	Single	0	0
12	Successful	1	Unmarried partner	0	1
13	Unsuccessful	0	Unmarried partner	0	1
14	Successful	1	Single	0	0
15	Unsuccessful	0	Unmarried partner	0	1
16	Unsuccessful	0	Single	0	0
17	Successful	1	Married	1	0
18	Unsuccessful	0	Single	0	0
19	Successful	1	Unmarried partner	0	1
20	Successful	1	Married	1	0
21	Successful	1	Unmarried partner	0	1
22	Unsuccessful	0	Single	0	0
23	Unsuccessful	0	Single	0	0
24	Unsuccessful	0	Unmarried partner	0	1
25	Unsuccessful	0	Married	1	0
26	Successful	1	Unmarried partner	0	1
27	Successful	1	Unmarried partner	0	1
28	Successful	1	Unmarried partner	0	1
29	Successful	1	Married	1	0
30	Unsuccessful	0	Single	0	0
31	Successful	1	Married	1	0
32	Successful	1	Unmarried partner	0	1
33	Unsuccessful	0	Unmarried partner	0	1
34	Successful	1	Unmarried partner	0	1

Person ID	Cessation (Y)	Cessation (Y)	Intimate Relationship (X_3)	Dummy Variables for Intimate Relationship	
				"Married" Dummy (X_{3a})	"Partner" Dummy (X_{3b})
35	Unsuccessful	0	Unmarried partner	0	1
36	Successful	1	Married	1	0
37	Unsuccessful	0	Unmarried partner	0	1
38	Unsuccessful	0	Unmarried partner	0	1
39	Successful	1	Married	1	0
40	Successful	1	Single	0	0
41	Unsuccessful	0	Unmarried partner	0	1
42	Successful	1	Single	0	0
43	Successful	1	Married	1	0
44	Successful	1	Married	1	0
45	Unsuccessful	0	Single	0	0
46	Successful	1	Unmarried partner	0	1
47	Unsuccessful	0	Married	1	0
48	Unsuccessful	0	Single	0	0
49	Unsuccessful	0	Single	0	0
50	Successful	1	Unmarried partner	0	1
51	Successful	1	Married	1	0
52	Successful	1	Married	1	0
53	Successful	1	Unmarried partner	0	1
54	Successful	1	Unmarried partner	0	1
55	Successful	1	Married	1	0
56	Unsuccessful	0	Married	1	0
57	Successful	1	Single	0	0
58	Unsuccessful	0	Single	0	0
59	Unsuccessful	0	Married	1	0
60	Unsuccessful	0	Married	1	0
61	Successful	1	Single	0	0
62	Unsuccessful	0	Single	0	0
63	Successful	1	Unmarried partner	0	1
64	Successful	1	Unmarried partner	0	1

(Continued)

Person ID	Cessation (Y)	Cessation (Y)	Intimate Relationship (X_3)	Dummy Variables for Intimate Relationship	
				"Married" Dummy (X_{3a})	"Partner" Dummy (X_{3b})
65	Unsuccessful	0	Single	0	0
66	Successful	1	Married	1	0
67	Unsuccessful	0	Married	1	0
68	Successful	1	Single	0	0
69	Successful	1	Unmarried partner	0	1
70	Successful	1	Unmarried partner	0	1
71	Unsuccessful	0	Single	0	0
72	Unsuccessful	0	Unmarried partner	0	1
73	Successful	1	Married	1	0
74	Unsuccessful	0	Unmarried partner	0	1
75	Successful	1	Married	1	0

The independent variables in her logistic regression analysis included number of past attempts (X_1), years of education (X_2), the married dummy (X_{3a}), and the unmarried partner dummy (X_{3b}). Collinearity is an issue of relationships among X's, so it remains useful to examine the correlations among independent variables in a logit analysis, just as for linear regression. The highest (in absolute value) correlation is –0.538 between the married dummy (X_{2a}) and the partner dummy (X_{2b}). As mentioned before, it is often the case that dummies representing categories of the same independent variable are rather highly correlated, and no other correlations are so large as to suggest any collinearity problems.

The following is software output for the logistic regression analysis of cessation (Y), with independent variables past attempts (X_1), education (X_2), and the married (X_{3a}) and unmarried partner (X_{3b}) dummies:

Pseudo-R^2 = 0.236

Log-likelihood = –41.3625

	b	SE of b	z	Wald	p-value
Intercept (constant)	-2.422	1.345	-1.801	3.244	0.072
Past attempts (X_1)	-0.132	0.060	-2.190	4.795	0.029
Education (X_2)	0.141	0.071	1.994	3.974	0.046

	b	SE of b	z	Wald	p-value
"Married" dummy (X_{3a})	1.380	0.722	1.910	3.648	0.056
"Partner" dummy (X_{3b})	1.439	0.640	2.247	5.048	0.025

This pseudo-R^2 is 0.236; if it were the actual R^2 in a linear regression, that might not sound so impressive. However, remember that with logistic regression, one can no longer say that 23.6% of variance in cessation (Y) is explained by this set of X's. Because pseudo-R^2 is not quite the same as the R^2 in linear regression, and different versions of pseudo-R^2 can differ for the same model, it is best to interpret this as a very approximate rather than precise measure of model fit, with limited utility as a stand-alone figure. (The software used here provided both Cox and Snell's pseudo-R^2, shown above, and Nagelkerke's, which had a value of 0.316.)

Another way to examine model fit is via the 2 × 2 classification table. The following classification table indicates that 33 cases for which the estimated $P\{Y = 1\} > 0.50$ actually had Y = 1. Also, 23 cases for which the estimated $P\{Y = 1\} < 0.50$ actually had Y = 0. This means that in 75% of cases (56 out of 75) the model "got it right" by this standard. However, use of the classification table to assess the model fit also has shortcomings, as discussed earlier. For instance, suppose that instead of running the logistic regression model, we simply used the observed proportion of successes (42/75 = 0.56) as an estimate of P{Y = 1} for every case in the sample. Then we would be "correct" in 42 of the cases, or 56%, which is not so much less than the percentage obtained from running the model. So again, this assessment of model fit is rather crude. Note that some software produces a classification table based on having calculated the estimated probability of {Y = 1} for each case from b's obtained when excluding that case from the data. This can produce minor differences in the displayed table.

	Estimated P{Y = 1}	
Observed Y	> 0.50	< 0.50
1	33 (Correct)	9 (Incorrect)
0	10 (Incorrect)	23 (Correct)

To assess the statistical significance of numerical independent variables in logistic regression, we can still use the p-values in the output, now based on z-statistics rather than the t-statistics used in linear regression. The effect of past attempts (X_1) on cessation (Y)—or more precisely on cessation log odds, odds, or probability—is statistically significant, controlling for education (X_2) and intimate relationship status (X_3). Education (X_2) also has a statistically significant effect on cessation (Y), holding other X's constant. Note the additional column labeled "Wald," in which the values are the squares of the z-statistics given here. Software output usually includes just one column of z or Wald, not both, because they convey the same information. (The two columns in the table here actually came from two different software packages.) For the Wald values, the p-value is obtained from a chi-square table with 1 df. However, a chi-square distribution with 1 df is defined as the distribution of a squared z-value, and the tail area beyond z^2 under the 1 df chi-square curve will be the same as the combined tail areas beyond z and –z under the standard normal curve. So the same (two-sided) p-value will be obtained either way. Reported values may slightly differ across software packages, due to computational choices such as rounding, but any discrepancy will be very minor.

Because the intimate relationship status variable (X_3) required more than one dummy, a comparison of nested models (in particular, models with and without the two dummy variables representing intimate relationship status) is required in order to make an overall assessment of statistical significance. As discussed earlier, for a model like logistic regression that is estimated via maximum likelihood, the analyst will compare nested models via a chi-square test rather than the F-test used in linear regression and obtain the p-value by looking up the calculated chi-square statistic in a chi-square table (page 247).

The chi-square statistic requires the log-likelihood given in the output for the model with all the X's (log-likelihood$_1$) as well as that from the logistic regression that excludes the dummy variables being tested (log-likelihood$_0$). Chi-square is calculated as (–2 × log-likelihood$_0$) – (–2 × log-likelihood$_1$), again taking care in case our statistical software output shows –2 log-likelihood instead of log-likelihood. The output for the model with all the X's shows that log-likelihood$_1$ is –41.3625, and hence –2 × log-likelihood$_1$ is 82.725. From output—not shown here—for a logistic regression that leaves out the two intimate relationship status dummies (that is, including only the past attempts (X_1) and education (X_2) variables), log-likelihood$_0$ = –44.4315, and –2 × log-likelihood$_0$ = 88.863. The chi-square statistic for this test is therefore (–2 × log-likelihood$_0$) – (–2 × log-likelihood$_1$) = 88.863 – 82.725 = 6.138.

Looking up the p-value in a chi-square table requires the appropriate degrees of freedom. The degrees of freedom for this chi-square test are the difference in the number of independent variables between the two nested models (or more generally, the difference in the number of parameters each model is estimating). Here df is 4 – 2 = 2 (or equivalently 5 – 3 = 2, if we include the intercept among the parameters being estimated). The chi-square table shows that the calculated statistic of 6.138, with 2 df, implies a p-value lying between 0.025 and 0.05. (Chi-square p-value calculators online or in software provide a more precise p-value of 0.046.) This p-value is less than the 0.05 cutoff, so the regression with the two intimate relationship status dummies does indeed fit significantly better than the regression without them. The program director interprets this as showing that there is at least some statistically significant difference in estimated cessation log odds, odds, and probabilities across intimate relationship status (X_3) categories, controlling for past attempts (X_1) and education (X_2).

As with dummy variables in linear regression, p-values from z-statistics for the individual dummy variables are helpful in indicating whether there is a statistically significant difference in estimated cessation odds between a particular dummy category and the reference category. The p-value from the z-statistic for the married dummy is 0.056, indicating a marginally statistically significant difference in cessation (Y) between married and single participants, holding past attempts (X_1) and education (X_2) constant. The difference in estimated cessation log odds, odds, or probabilities between unmarried partnered participants and single participants is statistically significant at the 0.05 level, as the p-value for the partner dummy is 0.025.

To examine the statistical significance of the difference between two dummy categories (married and unmarried partner in this case), the program director can switch the reference category to either married or partner, and rerun the logistic regression analysis. Choosing married as the reference category, and rerunning the analysis with partner and single dummies, the p-value for the partner dummy is 0.929. This indicates that there is no statistically significant difference in estimated cessation log odds, odds, or probabilities between unmarried partnered participants and married participants, holding other independent variables constant. In sum, the program director can discern a difference between single and married participants, and perhaps between single and unmarried partnered participants, depending on how the borderline statistical significance is treated. But there is not much evidence of a difference between married and unmarried partnered participants.

When interpreting b for logistic regression, it is important to keep in mind that the logit model is written for the log odds of {Y = 1} (here, log odds of successful cessation). For a numerical X, b shows the estimated change in the log odds when X increases by one unit, holding other X's constant. If X is a dummy, then b is the estimated difference in log odds between the dummy's named category and the reference, again holding other independent variables constant. However, with the inherent difficulty in understanding a change or difference in log odds, it is better to focus on the odds ratio e^b and its interpretation as an estimated multiplicative effect on odds. The following table adds the odds ratios e^b for all of the X's to the previous table of logistic regression results:

	b	SE of b	z	Wald	p-value	e^b
Intercept (constant)	-2.422	1.345	-1.801	3.244	0.072	0.089
Past attempts (X_1)	-0.132	0.060	-2.190	4.795	0.029	0.876
Education (X_2)	0.141	0.071	1.994	3.974	0.046	1.151
"Married" dummy (X_{3a})	1.380	0.722	1.910	3.648	0.056	3.975
"Partner" dummy (X_{3b})	1.439	0.640	2.247	5.048	0.025	4.216

For the numerical X's—here past attempts (X_1) and education (X_2)—estimated odds of {Y = 1} are multiplied by e^b as X increases by one unit, holding other independent variables constant. Therefore, estimated odds of successful cessation {Y = 1} are multiplied by $e^{b_1} = 0.876$ for each additional past attempt (X_1); in other words, estimated odds of successful cessation {Y = 1} decrease 12.4% as past attempts (X_1) increase by 1, holding other X's constant. Estimated odds of successful cessation {Y = 1} are multiplied by $e^{b_2} = 1.151$ as education (X_1) increases by 1 year. Alternatively, estimated odds of successful cessation {Y = 1} increase 15.1% for each additional year of education that a participant has, holding other X's constant. In this sample, past attempts (X_1) had mean 5.427 and standard deviation 4.788, and education (X_2) had mean 17.107 and standard deviation 4.112. As these descriptive statistics suggest that an increase of one unit is not an unduly large change in either past attempts (X_1) or education (X_2), the estimated changes in cessation odds suggest that the effects of these variables are large in real terms.

Holding other independent variables constant, the odds of cessation for married participants are estimated to be 3.975 times those of single participants. This is an extremely large difference in estimated cessation odds between these two intimate relationship status categories. This could be put in percentage terms as before, estimating 297.5% greater odds of cessation for married than for single participants. However, when a percentage like this becomes so large, people may not have much intuitive sense of its meaning anyway, and it therefore may be more helpful to simply leave the figure in its multiplicative form. The odds ratio comparing estimated cessation odds between unmarried partnered participants and single participants is also extremely large (4.216). When switching the reference category to married, the b for the partner dummy is 0.059, which is also the difference in these categories' b's in the results above (1.439 – 1.380 = 0.059). $e^{0.059}$ is 1.061, meaning that estimated odds of cessation for unmarried partnered participants are 1.061 times greater, or 6.1% more, than for married participants. (This figure could also be obtained by dividing the previous two odds ratios: 4.216/3.975 = 1.061.) However, recall that this difference is not statistically significant ($p = 0.929$).

The program director also wanted to report some results using probabilities. For those interpretations, she used the strategy of setting other independent variables at their mean or modal values while focusing on changes in a single independent variable. In this sample, means for number of past attempts (X_1) and education (X_2) were 5.427 and 17.107, respectively. Also, the modal intimate relationship status (X_3) category was that of unmarried partner.

For intimate relationship status (X_3), she calculated the following probabilities of successful cessation for an otherwise "typical" participant (that is, at the sample mean on both the past attempts (X_1) and the education (X_2) variables). The estimated logistic regression equation was $L = -2.422 - 0.132\ X_1 + 0.141\ X_2 + 1.380\ X_{3a} + 1.439\ X_{3b}$. Plugging in these means, and the 0 or 1 values of the dummies for the different relationship statuses, yields $L = -2.422 - 0.132\ (5.427) + 0.141\ (17.107) + 1.380\ (1) + 1.439\ (0) = 0.654$ for married participants, $L = -2.422 - 0.132\ (5.427) + 0.141\ (17.107) + 1.380\ (0) + 1.439\ (1) = 0.713$ for unmarried partnered participants, and $L = -2.422 - 0.132\ (5.427) + 0.141\ (17.107) + 1.380\ (0) + 1.439\ (0) = -0.726$ for single participants. Values of e^L are 1.923, 2.040, and 0.484, respectively. Using the probability formula $P\{Y = 1\} = \frac{e^L}{1 + e^L}$, these correspond to estimated cessation probabilities of 0.66 for married participants, 0.67 for unmarried partnered participants, and 0.33 for single participants. Single participants seem to have a substantially lower probability of cessation than those in the other relationship categories.

She next produced a graph (Figure 7.4) showing how the estimated probability of cessation changes with the number of past attempts, with education set at its mean (17.107) and intimate relationship status set at its mode (unmarried partner). A variety of options for such graphs are available in most statistical software packages. Here she used the observed range of past attempts in the sample as the range for the horizontal axis.

We know that the estimated probability will not be a linear function of the past attempts, but this looks pretty close to a straight line. That is because none of these probabilities are too high or too low, and the range of the number of past attempts is not so large. (The number of past attempts cannot be a decimal value, of course, but the dots marking the estimated probabilities for each whole number value are connected to help with the visualization.) Changes in the estimated P{Y = 1} from one value of past attempts to the next are not all the same, so formally this is not a straight line, but these changes only range from –0.033 to –0.021, which may not be enough to make them look very different relative to the scale of the graph's vertical axis. In any case, it seems clear that the number of past attempts has a meaningful effect on the probability of cessation.

She produced a similar graph for years of education, again setting other independent variables at their mean or mode, and using the sample range of education (Figure 7.5).

FIGURE 7.4 ● Estimated Cessation Probability for "Typical" Case by Past Attempts to Quit

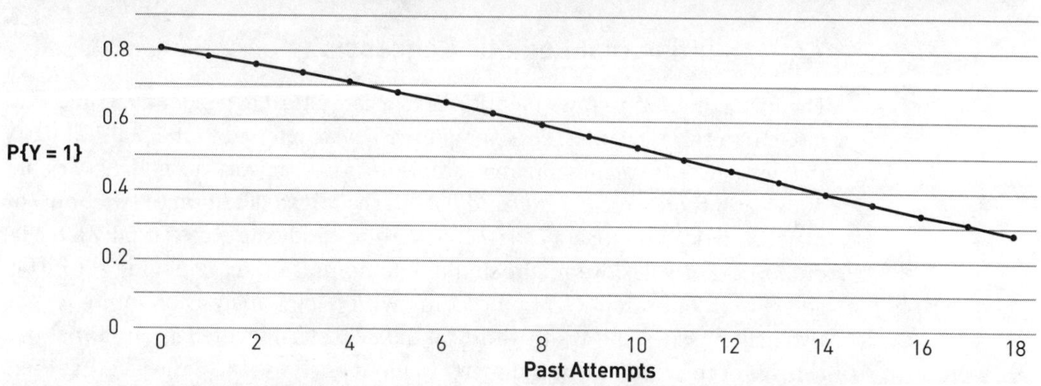

FIGURE 7.5 ● Estimated Cessation Probability for "Typical" Case by Years of Education

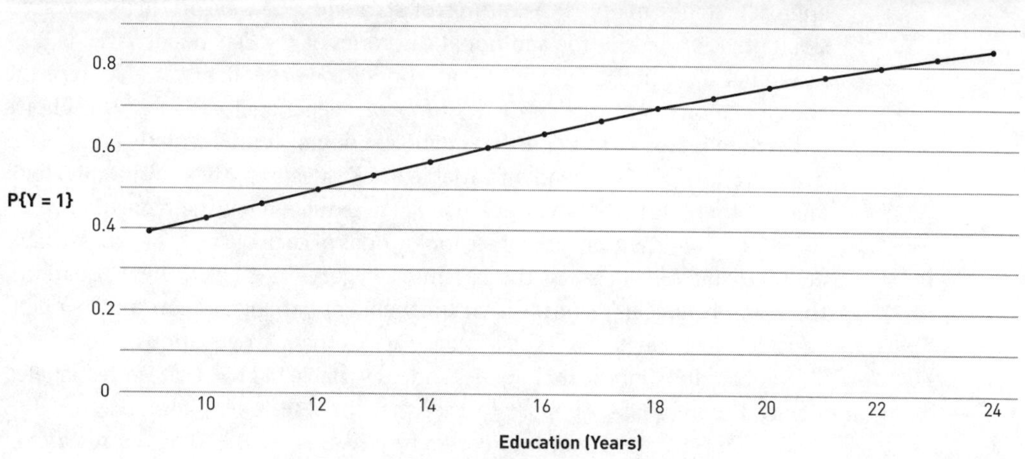

The nonlinear nature of the relationship may be slightly more apparent in this plot, but still it looks rather close to a straight line. Changes in the estimated probability for adjacent years of education range from 0.020 to 0.035, so again this is not so far off from linearity. As with past attempts, this display of cessation probabilities indicates that education has an important effect.

7.4 Some Extensions of Logistic Regression

Dichotomous dependent variables are very common in social science research, and the logistic regression framework discussed in this chapter is an effective tool for analyzing such data. There are also some important extensions of this framework, two of which we will note briefly here: multinomial logistic regression, for a dependent variable with more than two categories, and ordinal logistic regression, used when those categories are inherently ordered. We also introduce probit regression, an alternative method for a dichotomous dependent variable that is popular in some fields. Advanced texts give more detail on these methods, so readers interested in following up on our brief introduction can learn more there.

7.4.1 Multinomial Logistic Regression

The first extension that we mention is to a categorical dependent variable with more than two categories. For example, earlier we referred to the political party that respondents voted for as an example of a two-category dependent variable. But what if there were a third candidate in the race, in addition to two from the major American political parties? If that third candidate received only a trivial proportion of votes, we might simply ignore him or her, removing his or her voters from our sample and proceeding with a logit analysis of voting for the two major parties. But if this third candidate instead secured a substantial percentage of the votes, we would prefer that our analysis reflect the fact that there were three candidates, rather than being restricted to the two major parties.

A categorical dependent variable that has three or more categories may be analyzed using *multinomial logistic regression*, or multinomial logit for short. This approach shares many core features of standard two-category logistic regression but is adapted for the additional categories of the dependent variable. The researcher first chooses a baseline category of the dependent variable. Like the choice of a reference category with dummy independent variables, this choice of baseline is arbitrary, and different choices do not fundamentally change the overall results. If the dependent variable has K categories, the multinomial logit analysis then produces what look like $K - 1$ estimated logistic regression equations. However, these equations are for log odds of each of the $K - 1$ other dependent variable categories vs. the baseline category. In each of these equations, there will be a coefficient for each independent variable, and an independent variable's coefficients will typically vary across the $K - 1$ equations.

To illustrate, suppose that our dependent variable has $K = 4$ categories, labeled simply 1, 2, 3, and 4. Suppose further that the researcher chose category 4 as the baseline for their multinomial logistic regression and that the researcher decided to use three independent variables in the regression. Then their analysis produces—all simultaneously in one estimation process, not through running a series of separate regressions—a set of three logit-type estimated equations. These may be written as

$$\log\frac{P\{Y = 1\}}{P\{Y = 4\}} = a^{(1)} + b_1^{(1)}X_1 + b_2^{(1)}X_2 + b_3^{(1)}X_3,$$
$$\log\frac{P\{Y = 2\}}{P\{Y = 4\}} = a^{(2)} + b_1^{(2)}X_1 + b_2^{(2)}X_2 + b_3^{(2)}X_3, \text{ and}$$
$$\log\frac{P\{Y = 3\}}{P\{Y = 4\}} = a^{(3)} + b_1^{(3)}X_1 + b_2^{(3)}X_2 + b_3^{(3)}X_3.$$

The equations report on the relative odds of Y = 1, 2, or 3 vs. Y = 4. The superscripts (1), (2), and (3) on the a's and b's are necessary because these coefficients will in general differ across these equations. That is, an independent variable's estimated effect on the relative odds of {Y = 1} vs. {Y = 4} will probably not be the same as its effect on the odds of {Y = 2} vs. {Y = 4}. This means that a multinomial logit analysis could, depending on how many categories Y has and how many independent variables are being used, generate quite a large amount of output for the researcher to sort through.

If we wanted to see the estimated equation for {Y = 1} vs. {Y = 2}, say, we could repeat the analysis with {Y = 2} as the baseline; again, the choice of baseline does not fundamentally change results, so differences are merely cosmetic. Alternatively, we could think about this in light of our rules for logs in Chapter 5. By those rules, $\log\frac{P\{Y = 1\}}{P\{Y = 2\}} = \log P\{Y = 1\} - \log P\{Y = 2\}$. Similarly, $\log\frac{P\{Y = 1\}}{P\{Y = 4\}} = \log P\{Y = 1\} - \log P\{Y = 4\}$ and $\log\frac{P\{Y = 2\}}{P\{Y = 4\}} = \log P\{Y = 2\} - \log P\{Y = 4\}$. This means that $\log\frac{P\{Y = 1\}}{P\{Y = 2\}} = \log\frac{P\{Y = 1\}}{P\{Y = 4\}} - \log\frac{P\{Y = 2\}}{P\{Y = 4\}}$. So if we want the estimated coefficients for the equation of {Y = 1} vs. {Y = 2}, we can obtain these by taking the estimated coefficients in the equation for {Y = 1} vs. {Y = 4} and subtracting the estimated coefficients in equation for {Y = 2} vs. {Y = 4}. That operation will exactly match the results that we would obtain in the equation for {Y = 1} vs. {Y = 2} when using {Y = 2} as the baseline. Because all of the equations are being estimated at the same time, rather than in separate steps, results will be consistent in this respect no matter what baseline category we choose.

Otherwise, much that we discussed for two-category logistic regression above remains the same. Assessing statistical significance for a single variable in one of the equations is no different, and we can still use –2 × log-likelihood to compare nested models. We just need to be careful in counting degrees of freedom in that case, as adding or removing a single independent variable means adding or removing it in each of the equations, which will be a change of more than one df. Interpretations of b are also like those in two-category logit; our main additional challenge in reporting is to be careful that we are clear on which odds (that is, which category of Y vs. which other category of Y) a particular e^b is addressing.

7.4.2 Ordinal Logistic Regression

Sometimes the categories of a categorical dependent variable are inherently ordered. For instance, a survey question may ask respondents if they strongly agree, agree, disagree, or strongly disagree with some statement. If we were to analyze this agreement as a four-category dependent variable, we could simply apply the multinomial logit approach just mentioned. However, researchers often try to take advantage of a dependent variable's ordinal nature to simplify the model. There are some different ways to do that; here we discuss the "cumulative logit" approach to *ordinal logistic regression* (or *ordered logistic regression*).

In that example, there would be three ways to lump categories into broader "agree" and "disagree" groups while still respecting the categories' inherent order. These would be (a) {strongly agree} vs. {agree, disagree, strongly disagree}; (b) {strongly agree, agree} vs. {disagree, strongly disagree}; and (c) {strongly agree, agree, disagree} vs. {strongly disagree}. Each of these lumpings of categories has a "more agreement" side and a "more disagreement" side, with the constituent categories changing depending on where we made the cut in the original list of four ordered categories.

The simplification introduced in ordinal logit is that each independent variable is assumed to have the same effect on the odds of "more agreement" vs. "more disagreement" no matter where we cut the categories. We cannot extend that to the intercept, because the sheer number of cases that fall on the agree or disagree side of the cut will change depending on which cut we are considering. But because we will not have distinct b's for each cut, there will be much less output to interpret in ordinal logit than there would be in a multinomial logit analysis of the same data. In our example, we could write the following equations for a model with three independent variables, where SA means "strongly agree," and so on:

$$\log\frac{P\{SA\}}{P\{A,\ D,\ SD\}} = a^{(1)} + b_1X_1 + b_2X_2 + b_3X_3,$$

$$\log\frac{P\{SA,\ A\}}{P\{D,\ SD\}} = a^{(2)} + b_1X_1 + b_2X_2 + b_3X_3, \text{ and}$$

$$\log\frac{P\{SA,\ A,\ D\}}{P\{SD\}} = a^{(3)} + b_1X_1 + b_2X_2 + b_3X_3.$$

The intercepts $a^{(1)}$, $a^{(2)}$, and $a^{(3)}$ change across these equations, but the b's do not. As with multinomial logit, these are all estimated together, rather than in separate steps.

One practical issue in using ordinal logit is that we must make sure we understand how our software is assigning categories to the numerator and denominator of the log odds expression, so that we do not misunderstand the meaning of b's sign or our calculated e^b. It is natural to assume that a positive b means that higher values of that X are associated with greater odds of a "high" rather than "low" response on the dependent variable, but it is important to verify that this is how our software is reporting results, especially as formal presentations of the model can be quite confusing in this respect. Aside from this concern, however, logit-type interpretations apply, and we can assess statistical significance or compare nested models in the expected ways.

An important question is whether the simplification of allowing the same b's to apply in each of the implied logistic regression equations is warranted. That is, maybe this simplification is too extreme, and different b's are required after all to describe our data more accurately. Many software packages allow for an explicit test of this simplification, producing a p-value for the hypothesis that these effects are identical across the different cuts. If the simplification in fact seems unwarranted in our data, then we can revert to use of the multinomial logit model, which does not rely on any ordering of the categories of the dependent variable. Although the simplification of the ordinal logit approach is welcome for its reduction of the complexity of multinomial logit results, it will not be appropriate in all data contexts.

7.4.3 Probit Regression

In many social science fields, logistic regression is the most common approach for the analysis of a dichotomous dependent variable. However, in these fields one sometimes also encounters *probit regression* for a dichotomous dependent variable, and probit has greater popularity in disciplines such as political science and economics. Recalling the original problem of trying to carry out linear regression for a 0–1 dependent variable, probit offers a different solution than logit. As in logit, probit involves a linear regression for some quantity that then can be transformed into a probability, but probit does this via a different quantity and transformation. Rather than working with odds and log odds, probit instead considers the probability of $\{Y = 1\}$ through the standard normal distribution.

In particular, it estimates a model of the form $Z = a + b_1 X_1 + b_2 X_2 + \ldots$, where the implied $P\{Y = 1\}$ is found by the area under the standard normal curve, from $-\infty$ to Z. Software can obtain probit model estimates by maximum likelihood, as it does with logit, but the different form of the model means that there is a different interpretation of b than that we saw previously for logit. Our core interpretation of b in logit was as a change in the log odds, whereas for probit the core interpretation of b is as a change in the Z-score in the standard normal distribution. As in logit, great care must be made when attempting to extend the interpretation of b to discuss changes in $P\{Y = 1\}$, because the same change in the Z-score does not always imply the same change in probability.

For example, if our probit results show that the predicted Z-score is 0.25 greater for men than for women, we cannot use that to make a general statement of how the probability differs for men and women, as that difference will depend on the values of all the other independent variables too. Suppose that for a particular configuration of all the other independent variables, the predicted Z is 1.37 for men and 1.12 for women. The standard normal table then tells us that the predicted probabilities are 0.915 for men and 0.869 for women, a difference of 0.046. But suppose that for another set of values for the other independent variables, we have a predicted Z of −0.46 for men and −0.71 for women. In that case, the predicted probabilities are 0.323 for men and 0.239 for women, a difference of 0.084. The difference in probabilities implied by a given difference in Z will change depending on what part of the standard normal curve is involved. More background on this and other aspects of probit regression, including its similarities and differences with logistic regression, can be found in advanced texts.

7.5 Exercises

7.5.1 Exercise 1: Clearance of Violent Crimes by Arrest

For a final class project, a group of students decided to use 60 randomly selected violent crime incidents from the FBI's National Incident-Based Reporting System (NIBRS) data to examine factors that influence crime clearance outcomes. A crime is considered cleared if a suspect was arrested for the crime; the students created a dichotomous variable that indicated whether a crime incident was cleared by arrest or not within a year. Based on the victim-devaluing perspective, students identified three extralegal characteristics of the victim that might influence the clearance outcome: age (in years), gender (male or female), and race (white or non-white). The victim-devaluing perspective posits that crime incidents with older, male, and white victims are more likely to be cleared than those with younger, female, and non-white victims, with these differences attributable to differences in the social standing of different kinds of victims. The students also included victim–offender relationship, because it is typically easier to clear crime involving an offender known to the victim, though there will not necessarily be an arrest in all such cases. Victim–offender relationship included three categories: offender known to the victim (including family, intimates, friends, and acquaintances), stranger, and unknown relationship.

The data are as follows:

Data:

Incident ID	Clearance Outcome	Victim's Age	Victim's Gender	Victim's Race	V–O Relationship
1	Cleared	40	Female	Non-white	Unknown
2	Not cleared	31	Female	Non-white	Offender known
3	Not cleared	11	Male	Non-white	Unknown
4	Cleared	30	Female	white	Offender known
5	Not cleared	13	Female	white	Unknown
6	Not cleared	31	Female	white	Unknown
7	Cleared	58	Male	white	Offender known
8	Cleared	58	Female	white	Stranger
9	Not cleared	35	Male	white	Unknown
10	Not cleared	42	Male	Non-white	Unknown
11	Not cleared	41	Male	white	Stranger
12	Not cleared	24	Male	Non-white	Offender known
13	Not cleared	21	Male	white	Stranger
14	Cleared	27	Male	white	Stranger

Incident ID	Clearance Outcome	Victim's Age	Victim's Gender	Victim's Race	V–O Relationship
15	Cleared	27	Female	white	Stranger
16	Not cleared	52	Male	white	Unknown
17	Not cleared	34	Male	white	Stranger
18	Not cleared	5	Female	white	Offender known
19	Not cleared	29	Female	Non-white	Offender known
20	Not cleared	35	Male	Non-white	Unknown
21	Cleared	40	Male	Non-white	Unknown
22	Not cleared	14	Female	white	Offender known
23	Not cleared	38	Male	white	Stranger
24	Cleared	37	Male	white	Offender known
25	Not cleared	23	Male	white	Offender known
26	Cleared	44	Female	Non-white	Offender known
27	Not cleared	38	Male	Non-white	Unknown
28	Not cleared	51	Male	Non-white	Unknown
29	Not cleared	34	Female	white	Offender known
30	Not cleared	22	Male	white	Stranger
31	Cleared	24	Female	Non-white	Offender known
32	Cleared	36	Male	white	Unknown
33	Not cleared	23	Female	Non-white	Offender known
34	Cleared	37	Female	Non-white	Offender known
35	Cleared	59	Male	white	Offender known
36	Not cleared	41	Female	white	Offender known
37	Cleared	9	Male	Non-white	Offender known
38	Not cleared	14	Female	Non-white	Offender known
39	Not cleared	59	Male	Non-white	Offender known
40	Cleared	37	Female	Non-white	Offender known
41	Cleared	20	Female	white	Unknown
42	Cleared	21	Female	white	Offender known
43	Cleared	32	Female	white	Offender known
44	Not cleared	10	Female	Non-white	Offender known

(Continued)

Incident ID	Clearance Outcome	Victim's Age	Victim's Gender	Victim's Race	V–O Relationship
45	Not cleared	44	Male	white	Stranger
46	Not cleared	20	Male	Non-white	Offender known
47	Not cleared	20	Female	Non-white	Offender known
48	Not cleared	60	Male	Non-white	Offender known
49	Cleared	25	Female	white	Offender known
50	Cleared	37	Female	white	Offender known
51	Not cleared	6	Male	Non-white	Offender known
52	Not cleared	14	Female	white	Offender known
53	Not cleared	28	Male	white	Unknown
54	Not cleared	7	Female	white	Offender known
55	Cleared	23	Female	white	Offender known
56	Not cleared	50	Male	Non-white	Stranger
57	Cleared	33	Female	white	Offender known
58	Cleared	42	Male	Non-white	Unknown
59	Not cleared	16	Male	Non-white	Unknown
60	Cleared	31	Female	white	Offender known

1. Analyze the data in a multiple regression analysis that accommodates the dichotomous dependent variable (crime clearance).
2. Interpret the results of the analysis in Question 1.
3. Calculate the estimated probability of clearance for a crime incident in which a 39-year-old white man was victimized by a stranger.
4. Use your software to plot the estimated relationship between probability of clearance and victim's age, setting other independent variables at their mode.

7.5.2 Exercise 2

1. Suppose X is a dummy independent variable for "Foreign born" (with reference "U.S. born"), and the associated b was statistically significant. Give the proper interpretation of the following results in a logit model for "agree" vs. "disagree":

   (a) b = 0.078

   (b) b = –0.025.

2. Suppose X is a numerical independent variable for years living at current residence, and the associated b was statistically significant. Give the proper interpretation of the following results in a logit model for "agree" vs. "disagree":

   (a) b = 0.12

   (b) b = −0.07.

3. Suppose X represents hours spent studying, and b = 0.096 (with statistical significance) in a logistic regression model for "pass" (1) vs. "fail" (0). What happens to the estimated odds of passing if hours spent on studying per week increase by 7?

7.5.3 Exercise 3: Voting on Property Tax Referendum

City residents voted on a referendum to increase property taxes to support greater public school funding. A researcher randomly selected 800 residents and asked their opinions (for or against) on the referendum. Reflecting a particular theory of influences on support for public school spending, these data were analyzed via the following logistic regression model, with independent variables home ownership (owner or renter), education (high school or less, or college graduate), local status (city native or raised elsewhere), number of children, and gender (female or male). Logistic regression results are as follows:

Log-likelihood = −540.637

	b	SEof b	z	Wald	p-value
Intercept (constant)	−0.056	0.202			
Owner	−0.376	0.179	−2.109	4.414	0.035
High school or less	−0.150	0.178	−0.843	0.711	0.399
City native	0.583	0.190	3.068	9.413	0.002
Number of children	0.474	0.183	2.588	6.708	0.010
Female	0.059	0.156	0.378	0.143	0.705

1. Interpret the results as fully as possible.
2. An alternative theory says that only education, local status, and gender matter for this kind of issue. In a logit model with just these independent variables, log-likelihood = −546.256. What evidence do the results provide about the relative performance of the two theories in modeling opinions on the referendum?

7.6 Appendix

The logit model, with three independent variables, is $\log\frac{P\{Y = 1\}}{1 - P\{Y = 1\}} = \alpha + \beta_1X_1 + \beta_2X_2 + \beta_3X_3$; for convenience we will shorten this to $\log\frac{P\{Y = 1\}}{1 - P\{Y = 1\}} = L$, where $L = \alpha + \beta_1X_1 + \beta_2X_2 + \beta_3X_3$. Applying the antilog to both sides of that shortened expression, we have $\frac{P\{Y = 1\}}{1 - P\{Y = 1\}} = e^L$. Next, we multiply both sides by $(1 - P\{Y = 1\})$, and then multiply through the parentheses on the right side:

$$P\{Y = 1\} = (1 - P\{Y = 1\})e^L = e^L - e^L P\{Y = 1\}.$$

From $P\{Y = 1\} = e^L - e^L P\{Y = 1\}$, we can group the terms involving $P\{Y = 1\}$ on the left: $P\{Y = 1\} + e^L P\{Y = 1\} = e^L$, or $P\{Y = 1\}(1 + e^L) = e^L$.

Finally, we divide by $(1 + e^L)$ in order to leave $P\{Y = 1\}$ by itself on the left-hand side, as $P\{Y = 1\} = \frac{e^L}{1 + e^L}$.

An alternative version of this final expression is created by taking one more step to divide both the numerator and the denominator of the right-hand side by e^L. Then the numerator is simply 1, and the denominator is $e^{-L} + 1$, because 1 divided by e^L is e^{-L}. That gives a final expression of $P\{Y = 1\} = 1/(e^{-L} + 1)$.

8

Count Dependent Variables

Poisson Regression

In Chapter 7, we learned how to analyze a categorical dependent variable in multiple regression. In Chapter 8, we will discuss common methods for regression analysis when the dependent variable is a count, rather than a continuous measure. By count, we mean a variable that can only take integer (whole number) values and cannot be negative. For example, a person's number of children is such a count: it cannot be negative, and there is no such thing as a fraction of a child. When a count dependent variable is one that typically takes on very large values, such as a major automaker's number of cars sold in the United States last year, it is usually fine to simply treat it as if it were a continuous dependent variable, analyzing it using the regression methods that we learned in previous chapters. But when a *count variable* takes on small or zero values, as for number of children, it is better to respect the special characteristics of this kind of dependent variable. We will examine *Poisson regression* methods that are designed for analyzing count data, followed by a full example and exercises in the end of the chapter. Also, we give a brief introduction to some extensions of Poisson regression, including log-rate models, negative binomial regression, and zero-inflated models.

8.1 Poisson Regression for Count Dependent Variables

Applying linear regression analysis when the dependent variable is a count variable raises some of the same concerns that arise with a dichotomous dependent variable. Count variables are those that can only take on values that are whole numbers (integers) and restricted to be zero or positive: 0, 1, 2, 3, 4, and so on. We typically use the term for a variable whose values are not extremely large. Count variables often involve individual-level data, such as the number of children in a family, the number of business trips per year for a worker, or the number of interactions with police that a juvenile had in some time period. However, count variables also could report aggregate-level information, such as the number of homicides in a large city in a month.

When the dependent variable in our analysis is a count, it is questionable whether the usual multiple linear regression is appropriate. The observed values of Y in the sample are all nonnegative integers, but the predicted Y from linear regression can of course be negative and will not be restricted to integer values. Further, there are technical assumptions in linear regression that will not be met when the dependent variable is a count. For these reasons, researchers typically prefer that a count dependent variable be analyzed with tools that are specifically adapted for it. As noted above, in practice this is more important when the counts involved tend to be rather small; a count variable for which values are quite large usually can be analyzed via conventional linear regression without much worry.

Poisson regression is a common approach to regression with a count dependent variable. Like logistic regression, the model does not aim to predict Y directly; instead, it holds that a case's observed value of Y follows a Poisson distribution, which is a well-known distribution for count variables. The model then reports the expected value, or mean, of this distribution for that case (denoted as $\mu^{(Y)}$; μ is a common symbol for a mean, and the superscript Y makes clear which variable is involved). Under this model, the logged expected value (mean) of Y is given by a linear equation: $\log \mu^{(Y)} = \alpha + \beta_1 X_1 + \beta_2 X_2 + \beta_3 X_3 + \ldots$. Even though the model can indicate either positive or negative values of $\log \mu^{(Y)}$, $\mu^{(Y)}$ will be always positive: under the definition of logs in Chapter 5, $e^{\log \mu^{(Y)}} = \mu^{(Y)}$, and any power of e will return a positive number.

Even though Y itself, as a count variable, must take only integer values, the expected value (mean) of Y ($\mu^{(Y)}$) does not need to be an integer. This kind of situation is familiar to everyone; take, for instance, the count variable number of children that we mentioned earlier. News stories frequently report a fractional value for the average number of children of American adults, though any individual's actual number of children of course will be an integer. So we can still accommodate the characteristics of a count variable Y and the Poisson distribution even if the model returns a decimal value for the expected value (mean) of Y ($\mu^{(Y)}$). This is one benefit of the model's focus on $\mu^{(Y)}$ rather than on Y directly.

As in logistic regression, Poisson regression output will include the log-likelihood, allowing us to compare nested models with the chi-square test based on differences in $-2 \times$ log-likelihood. Output may also include the *deviance*, which reflects a $-2 \times$ log-likelihood comparison between the model being analyzed and the "saturated" model in which each case's count is fit exactly (in other words, $Y = \mu^{(Y)}$ for each case). The difference in deviance between two nested models will equal the difference in the models' $-2 \times$ log-likelihood, so this is an equivalent method of comparing nested models. Assessment of statistical significance for the effect of a single variable is again based on a statistic derived from b divided by its estimated standard error, and the resulting p-value.

Count data also permit some measures of how well the model fit the observed data that are not available in logistic regression for a dichotomous dependent variable. These are based on the difference between the observed Y for each case and the expected Y estimated by the model. One measure of fit that is provided on most output is a chi-square statistic that uses each case's squared difference between its observed and estimated expected Y under the model; this is divided by the expected Y, and summed over all cases, as $\Sigma\frac{(Y - \mu^{(Y)})^2}{\mu^{(Y)}}$. (It would be a bit more precise to put a hat on $\mu^{(Y)}$, like $\widehat{\mu^{(Y)}}$, to reinforce that this is an estimate, but that is awkward with the superscript Y.) A test of the hypothesis that the model "fits" the data is obtained from this measure by looking it up in a chi-square table for df = N – (number of X's) – 1. A small p-value indicates that the model does not fit the data well, because we are rejecting the hypothesis that it does fit. The deviance can be used this way also (see advanced texts for more detail); in most applications, the value of the chi-square statistic will be at least fairly similar to the value of the deviance. If this assessment indicates that the model does not fit the data well, we can think about other independent variables that could help improve the fit, similar to our response to a low R^2 in linear regression. This situation may also lead us to investigate some of the extensions of Poisson regression discussed in Sections 8.3.2 and 8.3.3.

Our software will obtain estimates (a and the b's) for these parameters of the Poisson regression model via maximum likelihood, so that $\log \mu^{(Y)} = a + b_1X_1 + b_2X_2 + b_3X_3 + \ldots$ is estimated from our data. (Again, we are not putting a hat on the left-hand side for clarity, but this is of course an estimate, as indicated by the use of b's rather than β's in the expression.) This expression immediately reminds us of linear regression with a logged dependent variable in Chapter 5, and we can similarly use the antilog to rewrite the model for the expected value of Y rather than the log expected value, so that we estimate that $\mu^{(Y)} = e^{(a+b_1X_1+b_2X_2+b_3X_3+\ldots)}$. This can also be written as $e^a\ e^{b_1X_1}\ e^{b_2X_2}\ e^{b_3X_3} \ldots$. No matter if we get a positive or negative value when calculating $(a + b_1X_1 + b_2X_2 + b_3X_3 + \ldots)$ from the estimates a, b_1, b_2, b_3, ... and some set of X values, raising e to that power will always return a positive number for the expected value of Y.

We can go through steps like those for the case of logged Y in Chapter 5 (Section 5.3.1) to confirm that a one-unit increase in X_1 will multiply the expected value of Y by e^{b_1}. Interpretations therefore refer to multiplicative changes in the expected value of Y. For example, suppose that Y measures the number of books that a survey respondent read last year and X_1 measures education (in years). Suppose too that our output shows statistical significance for this numerical X_1, and gives the estimate $b_1 = 0.24$. Then our interpretation is that we estimate that for each additional year of education, holding other independent variables constant, the expected number of books read is multiplied by $e^{0.24} = 1.271$, or a 27.1% increase. If X_2 measures the respondent's number of children, and $b_2 = -0.09$ (again assuming statistical significance), then each additional child will, holding other independent variables constant, multiply the expected number of books read by $e^{-0.09} = 0.914$, or an 8.6% decrease.

If X is a categorical variable, then the expected value of Y for the dummy category is e^b times the expected value of Y for the reference category, controlling for other X's in the regression. Continuing with the example of the number of books that a respondent read last year as Y, suppose that X_3 is a dummy variable "female" from the two-category categorical variable gender (female and male, so male is the reference category). If b_3 is 0.09, our interpretation is that the expected number of books read last year for women is $e^{0.09} = 1.094$ times the expected number of books read for men, holding other X's constant. In other words, the expected number of books read last year is 9.4% more for women than for men, controlling for other X's. If b_3 is –0.12, the expected number of books read last year for women is estimated to be $e^{-0.12} = 0.887$ times the expected number of books read for men, holding other X's constant. In other words, the expected number of books read last year is 11.3% less for women than for men, controlling for other X's.

Assessment of real-world significance is similar to its assessment in Chapter 5 in analyses with logged Y as the dependent variable. As before, we may decide, based on the descriptive statistics for some X, that something other than a one-unit increase in that X should be used in evaluating its real-world significance. But changes in the expected Y will be expressed in multiplicative terms, so our use of descriptive statistics for Y is not the same as in assessing real-world significance in (unlogged) linear regression.

8.2 Example: Education and Desired Number of Children

This example will provide practice in dealing with a count dependent variable via Poisson regression. Despite their financial advantages, more educated people may end up having fewer children than the less educated, because they tend to delay having children and may have fewer in response to different career demands. A group of researchers wondered if education has a similar sort of relationship with people's desired, rather than actual, number of children. They collected a sample of 39 adults who have completed their formal education and asked the respondents for their highest academic degree (high school, bachelor's degree, or graduate degree) and desired number of children. In addition to education, the researchers wanted to examine gender differences in the desired number of children. It is also reasonable to suspect that how many siblings a person has will affect their desired number of children, so that information was collected as well. The research hypothesis, units of analysis, measurements of independent and dependent variables, and hypothetical raw data are as follows:

> *Research hypothesis:* Education and number of siblings influence people's desired number of children. There is a gender difference in people's desired number of children.
>
> *Units of analysis:* 39 adults who completed their formal education (individual-level data).

Measurements of independent and dependent variables:

Y: Desired number of children.

X_1: Academic degree: "High school," "Bachelor's degree," "Graduate degree."

X_2: Gender: "Female," "Male."

X_3: Number of siblings.

Data:

Subject ID	Ideal Number of Children (Y)	Academic Degree (X_1)	Gender (X_2)	Number of Siblings (X_3)
1	2	Bachelor's	Male	3
2	4	Graduate degree	Male	3
3	0	Bachelor's	Male	0
4	2	High school	Female	2
5	2	Graduate degree	Male	1
6	4	Bachelor's	Female	5
7	3	High school	Female	2
8	1	Graduate degree	Male	0
9	1	Bachelor's	Female	0
10	2	Bachelor's	Male	2
11	0	Graduate degree	Female	3
12	3	Graduate degree	Male	3
13	2	High school	Female	1
14	3	High school	Female	2
15	1	Bachelor's	Female	1
16	0	Graduate degree	Female	1
17	0	Bachelor's	Female	0
18	4	High school	Female	3
19	5	High school	Male	2
20	2	Bachelor's	Female	1
21	2	Bachelor's	Female	1
22	2	Bachelor's	Female	2

(Continued)

Subject ID	Ideal Number of Children (Y)	Academic Degree (X_1)	Gender (X_2)	Number of Siblings (X_3)
23	5	Bachelor's	Male	7
24	8	Bachelor's	Male	2
25	3	High school	Female	3
26	3	High school	Male	1
27	2	Graduate degree	Female	0
28	3	High school	Female	2
29	4	Bachelor's	Male	4
30	0	High school	Male	0
31	0	Bachelor's	Female	1
32	1	Graduate degree	Male	0
33	4	Bachelor's	Female	5
34	5	Bachelor's	Female	4
35	0	Graduate degree	Female	0
36	9	Bachelor's	Male	4
37	3	Bachelor's	Male	4
38	3	Graduate degree	Male	0
39	0	High school	Female	1

In this study, the dependent variable Y (desired number of children) is a count variable taking on nonnegative integer values from 0 to 9. Concerned that the usual linear regression analysis would not be appropriate in this setting, the researchers decided to use Poisson regression. Academic degree (X_1) and gender (X_2) are categorical variables, so the researchers created dummy variables. Dummy variables for "bachelor's degree" (X_{1a}) and "graduate degree" (X_{1b}) represented the three-category academic degree variable (X_1) (with "high school" as the reference category). For the two-category gender variable (X_2), the researchers selected "Female" as the reference category and created a "male" dummy variable (X_{2a}). The Poisson regression analysis for this study thus includes four independent variables: bachelor's dummy (X_{1a}), graduate dummy (X_{1b}), male dummy (X_{2a}), and number of siblings (X_3).

Because collinearity concerns relationships among X's, it should still be considered in a Poisson regression analysis. The highest correlation among any independent variables (in absolute value) is –0.544, between the "bachelor's" (X_{1a}) and "graduate" (X_{1b}) dummies. None of the other correlations are at all large, and a rather strong negative correlation within the set of academic degree dummies is not a concern, so these correlations do not suggest a potential problem of collinearity. The results from Poisson regression using the bachelor's dummy (X_{1a}), the graduate dummy (X_{1b}), the male dummy (X_{2a}), and the number of siblings (X_3) are shown in the following table.

Chi-square = 37.160
df = 34
Chi-square/df = 1.093
Deviance = 44.644
Log-likelihood = –67.4324

	b	SE of b	z	Wald	p-value	e^b
Intercept (constant)	0.374	0.224	1.667	2.780	0.095	1.454
Bachelor's (X_{1a})	–0.272	0.264	–1.032	1.066	0.302	0.762
Graduate degree (X_{1b})	–0.535	0.326	–1.643	2.700	0.100	0.586
Male (X_{2a})	0.485	0.216	2.245	5.042	0.025	1.624
Number of siblings (X_3)	0.236	0.060	3.946	15.572	< 0.001	1.266

For Poisson regression of count data, we can assess the model's fit to the sample data by testing H_0: model fits the observed data, against H_1: model does not fit the observed data. This test uses the chi-square statistic and df provided in the Poisson regression output. From the results, chi-square is 37.160, and df is calculated as N – number of X's – 1 = 39 – 4 – 1 = 34. In the chi-square table (page 247), the p-value shown for chi-square = 37.160 and df = 34 is > 0.25 (a more precise p-value is 0.326), well above the usual 0.05 cutoff. The researchers therefore failed to reject H_0 (model fits the data), and concluded that the model does seem to fit the observed data adequately.

The categorical academic degree variable (X_1) required more than one dummy, so a comparison of nested models (models with and without the bachelor's (X_{1a}) and graduate degree (X_{1b}) dummies) is required to make an overall assessment of its statistical significance. Because estimated b's in a Poisson model are obtained via maximum likelihood, we do not rely on an F-test but instead follow the same process used in logistic regression (Chapter 7). This requires calculation of a chi-square statistic from the difference in –2 × log-likelihood between the two nested models, (–2 × log-likelihood$_0$) – (–2 × log-likelihood$_1$), with a p-value obtained from a chi-square table. In this case, log-likelihood$_0$ is the log-likelihood from the model that excludes the academic degree dummies, and log-likelihood$_1$ is from the model with all the X's. From the output above, the model with all of the X's has log-likelihood$_1$ = –67.432, so –2 × log-likelihood$_1$ = 134.864. From the results (not shown here) for a Poisson regression that excludes the two academic degree dummy variables, leaving only the male dummy (X_{2a}) and number of siblings (X_3), log-likelihood$_0$ is –68.848 and –2 × log-likelihood$_0$

is 137.696. The chi-square is then calculated as (–2 × log-likelihood$_0$) – (–2 × log-likelihood$_1$) = 137.696 – 134.864 = 2.832.

An equivalent comparison of the models can be carried out with the deviance. The output here shows 44.644 as the deviance for Model 1 with the academic degree dummies. Output for Model 0 without the academic degree dummies (not shown) gives a deviance of 47.476. The difference of the deviance statistics for the two nested models is 47.476 – 44.644 = 2.832. As expected, this exactly agrees with the difference in –2 × log-likelihood for the two models, so either approach can be used.

With degrees of freedom given by the difference in the number of X's between nested models (here 4 – 2 = 2), or equivalently the difference in the total number of model parameters (5 – 3 = 2), we can look up the p-value in the chi-square table. For 2 df, the chi-square table shows that the p-value corresponding to a chi-square statistic of 2.832 lies between 0.10 and 0.25 (more precisely, the p-value is 0.243). This p-value is more than the 0.05 cutoff, so adding the two academic degree dummies did not significantly improve fit. The researchers concluded that their analysis did not find a statistically significant difference by academic degree (X_1) in the expected number of children desired, holding gender (X_2) and number of siblings (X_3) constant.

The gender variable (X_2) only requires one dummy—the male dummy (X_{2a})—so we can use that dummy's p-value on the output to assess the statistical significance of gender (X_2). We can also use the output p-value to check statistical significance for a numerical independent variable, in this case number of siblings (X_3). As discussed for logistic regression in Chapter 7, the table of results shows a column labeled "z" and another labeled "Wald." Each entry in the Wald column is simply the square of the usual (b/SE) in the z column, at least up to minor differences due to rounding, so the two columns give the same information and software output will likely include only one of them. They each lead to the same p-value, because of the connection between their reference distributions (standard normal for z, and chi-square with 1 df for Wald).

The male dummy's (X_{2a}) p-value is 0.025, suggesting that there is a statistically significant gender difference in the expected number of children desired ($\mu^{(Y)}$), controlling for academic degree (X_1) and number of siblings (X_3). Number of siblings (X_3) also has a statistically significant effect on the expected number of children desired, holding other X's constant. The researchers then proceeded with interpreting b's for the male dummy (X_{2a}) and number of siblings (X_3).

When interpreting b for Poisson regression, it is important to remember that the Poisson regression model is written for the logged expected (mean) value of Y—that is, log $\mu^{(Y)}$—so interpretation in terms of the expected value of Y ($\mu^{(Y)}$) requires that we calculate e^b. For a dummy variable, the expected value of Y for the named category is estimated to be e^b times that for reference category. Here b for the male dummy (X_{2a}) is 0.485, so e^b is 1.624, indicating that analysis of this sample estimates that the expected number of children desired by men is 1.624 times that desired by women (or the expected number of children desired for men is 62.4% greater than that for women), controlling for other X's in the model. For a numerical independent variable, the expected value of Y is multiplied by e^b for each additional unit of X. The output shows that the b for the number of siblings (X_3) is 0.236 and e^b is 1.266. Therefore, the researchers estimate that for each additional sibling (X_3) that a person has, holding other independent variables constant, the expected number of children desired is multiplied by e^b =1.266 (a 26.6% increase).

The estimated gender difference in number of children desired (62.4% more children expected to be desired for men than for women) and the estimated effect of number of siblings (26.6% more children expected to be desired for each additional sibling) both seem quite large in real terms. The researchers' original hypothesis was supported in part by the analysis of these sample data. They suspected that there would be a gender difference and an effect of number of siblings, and both of these were apparent in the results. For each variable, results indicated both statistical significance and a large enough estimated difference or effect to suggest real-world significance.

However, their hypothesis concerning education was not supported in this analysis, because statistically significant differences among the educational categories were not found. This may be due to the relatively small sample, as the estimates e^b for multiplicative differences between the two dummy categories and the reference (high school) were quite different from 1 (0.762 for bachelor's and 0.586 for graduate degree). With only 39 cases in the sample, and three educational categories, comparisons between two categories are based on very few individuals. So although the hypothesis of educational differences was not formally supported here, this still would bear further investigation in any larger data sets collected in the future.

8.3 Some Extensions of Poisson Regression

The Poisson regression framework for count dependent variables can be extended in various ways to address other situations that may be encountered in research. Here we will quickly examine just a few of these extensions: log-rate models for aggregate-level rates derived from counts, negative binomial regression for overdispersed count data, and zero-inflated count models. More details on these models and interpretations of results can be found in advanced texts.

8.3.1 Log-Rate Models

In earlier chapters we worked with some variables that were constructed as rates, such as city crime rates. Those crime rates were obtained as the number of crimes reported in some year in a city divided by the city's population. To make the resulting figures more interpretable, we then multiplied by, say, 100,000, so that the final variable was understood as measuring the city's annual number of crimes per 100,000 population. The core step, though, was dividing the number of crimes by the city population. That allowed us to reasonably compare crime data from cities of different size.

This sort of rate is often a re-expression of an underlying count variable, here the number of crimes. That count may tend to be relatively small, as would be the case if we focused specifically on homicide in various American cities in some year. Then we may not be comfortable using the usual city homicide rate, or the logged homicide rate, as the dependent variable in a linear regression. That straightforward analysis would not acknowledge that the underlying city homicide measure is a count. But neither would we want to conduct a Poisson regression using the cities' raw homicide counts as the dependent variable, because we know that larger cities will in general have more homicides in a given year than smaller cities, simply because their larger populations mean more people at risk for homicide victimization. Indeed, that was the whole point of constructing the rates in the first place.

An alternative that respects the count nature of the dependent variable homicide *and* recognizes the need to account for city population is the *log-rate model* version of Poisson regression. If $\mu^{(Y)}$ represents the city's expected homicide count and M represents the city population, then the expected homicide rate is $\mu^{(Y)} / M$; here we will not take the additional step of multiplying by 100,000. If logged, this would be $\log(\mu^{(Y)} / M)$, and, by the rules for logs we know that $\log(\mu^{(Y)} / M) = \log \mu^{(Y)} - \log M$.

In the usual Poisson regression model discussed in this chapter, with three independent variables $\log \mu^{(Y)} = \alpha + \beta_1 X_1 + \beta_2 X_2 + \beta_3 X_3$. (As before, we would replace the α by a and the β's by b's when referring to the estimated parameters from an analysis.) The log-rate version includes log M on the right-hand side but without a coefficient (or, equivalently, a coefficient that is set to 1):

$$\log \mu^{(Y)} = \log M + \alpha + \beta_1 X_1 + \beta_2 X_2 + \beta_3 X_3.$$

In this formulation, if we subtract log M from both sides of the equation, the left-hand side would become $\log \mu^{(Y)} - \log M$, which is the same as the logged rate given by $\log(\mu^{(Y)} / M)$. The log-rate version of Poisson regression is therefore $\log(\mu^{(Y)} / M) = \alpha + \beta_1 X_1 + \beta_2 X_2 + \beta_3 X_3$. This model recognizes Y as a count that can be analyzed via Poisson regression while also acknowledging that it is sensible to focus on the rate rather than on the raw count. e^{b_1} from our statistical software output is interpreted as the multiplicative effect of a one-unit increase in X_1 on the expected rate, holding other independent variables constant.

To run this model, our software must be able to correctly incorporate the log M term with its fixed coefficient 1 into the Poisson regression. If the software does permit this, the software documentation will refer to it as an *"exposure"* or *"offset"* term. Be careful to check whether the software expects you to identify this with a variable representing M—in our example, the city population—or with a variable representing log M, the logged city population. (Sometimes both options will be available; usually exposure refers to M, and offset to log M.)

Note that in some research contexts M (here, population) may be theorized to also have an impact that goes beyond this role of accounting for exposure. For example, our theory may suggest that the conditions of life in larger cities contribute to their having in general higher rates of homicide than smaller cities, not just larger raw homicide counts due to their larger populations. In that case, population should be included as one of the independent variables in the Poisson regression, rather than only being included as this exposure/offset term. Note that we would need to decide whether or not population should be logged when it is acting as an independent variable in the model, whereas it must be logged in its role defining the log-rate model.

8.3.2 Overdispersion and Negative Binomial Regression

One of the technical aspects of the Poisson distribution is that its mean and variance will be equal. However, when an actual count variable is analyzed in a Poisson regression model, it often seems that there is more variability in its observed counts than this property of the Poisson distribution would imply. In many analyses, this seems to hold even after accounting for the fact that different cases will have different expected values (and variances) because they have different values of the independent variables.

This situation is referred to as *overdispersion*, and a common explanation for it is that there are additional independent variables that should have been included in the Poisson regression but were not, producing variability (heterogeneity) between cases that is not being captured by the Poisson regression model. A poorly fitting Poisson regression, as indicated by the chi-square test described earlier, suggests this possibility. One common diagnostic for overdispersion is therefore based on the chi-square statistic. If the Poisson regression model fits the data, and if df are at least moderately large, then the chi-square statistic divided by the df should, roughly, not be too much more than 1. If this ratio is substantially more than one, then overdispersion is suggested. If so, the researcher can revisit theory and previous research to try to identify other variables that should be included in the regression, but there may not be any obvious choices for those. Or perhaps good candidate variables can be identified from theory or previous research but are simply unavailable in the researcher's data set, or would be impossible to measure due to practical or ethical concerns.

In such a case, one classical approach was to use the chi-square/df ratio in an ad hoc adjustment of standard errors for the estimated Poisson regression parameters, in hopes that hypothesis tests would then be less affected by overdispersion. However, the more contemporary response is to replace the Poisson regression model with the *negative binomial regression* model. Negative binomial regression extends Poisson regression by adding a model parameter that represents this overdispersion. There are different technical approaches to doing this, and software documentation and advanced textbooks can be consulted on these details. In any case, the idea is that with this additional parameter in the model, the regression is accounting for overdispersion when it calculates other results such as b's and standard errors for the independent variables.

Aside from the overdispersion parameter, interpretations are in general like those for Poisson regression, so much of the previous discussion still applies. One difference, however, is that the chi-square test described earlier for testing the model's fit does not carry over to negative binomial regression. Also, to be confident that a negative binomial regression is producing valid results, we will want to have a rather large sample (certainly at least several times larger than the samples used in the Poisson regression example and exercise here). Note that the log-rate models discussed in the previous section can be based on the negative binomial instead of the Poisson if overdispersion is apparent.

Along with being used to compare nested negative binomial regressions, the reported log-likelihood for a negative binomial regression can also be used to test whether there is statistically significant overdispersion. That is because Poisson and negative binomial models with the same set of independent variables are themselves nested; the $-2 \times$ log-likelihood difference between a Poisson regression and a negative binomial regression with the same independent variables is a 1 df chi-square test of the hypothesis that there is no overdispersion. (The 1 df comes from the additional overdispersion parameter in the negative binomial regression.) Statistical software output will also give a p-value for the z-statistic created by dividing the estimated overdispersion parameter by its standard error, as an alternative way to test for statistically significant overdispersion.

8.3.3 Zero-Inflated Models

Zero-inflated refers to an excess of 0 values in the count dependent variable relative to how many would be expected if the Poisson regression model were correct. Such an excess could simply reflect overdispersion, as described above, but zero-inflated models have a different explanation in mind. Fishing is the classic example used to illustrate this explanation. If the number of fish caught on some day is Y, a zero value for Y could have happened in one of two ways. The first is that the person went fishing but had no luck and therefore did not catch any fish. The second is that the person did not even go fishing that day and so had no possibility of catching any fish. Zeros of the first sort do not represent any particular problem for Poisson regression, as this kind of randomness in the observed Y is understood as part of working with the Poisson distribution for a count variable. However, the second type of zero is not so easily accommodated, so that situation would inflate the observed data's number of zeros in a way that Poisson regression is not equipped to handle.

Zero-inflated Poisson regression models conceive of a two-part process paralleling the fishing example. The first part involves a logit or similar model for the probability that a case had even the possibility of a nonzero value (in the fishing context, the probability of that person going fishing), while the second part is a Poisson regression model for the count if a nonzero count was possible (the number of fish caught by those who went fishing). Although these parts are conceptually distinct, parameters for each are estimated simultaneously when this model is implemented in statistical software. Software also typically permits the analyst to include different sets of independent variables for the two parts of the process; for example, a particular independent variable may be related to the decision or opportunity to go fishing or not but unrelated to the number of fish caught once fishing. Negative binomial regression can also be extended this way, so software packages may offer a zero-inflated negative binomial regression model along with the zero-inflated Poisson.

Zero-inflated analyses are most appropriate when the research setting provides a meaningful analogy to the fishing story. If the researcher can provide a reasonable explanation, grounded in existing research or theoretical understanding of the research question, for why certain cases would not even have the possibility of a nonzero count for the dependent variable, then the zero-inflated Poisson may be a compelling alternative to the standard Poisson. However, if the research setting does not offer that sort of explanation for zero inflation, it may be better to simply view the excess zeros as indicative of overdispersion. That suggests attempting to address overdispersion through negative binomial regression before considering a zero-inflated analysis.

8.4 Exercises

8.4.1 Exercise 1: Employment Status and Alcohol Consumption

A researcher believes that employment status is a crucial factor in predicting people's alcohol consumption, even among those who are making an active effort to curb their drinking. To test her research hypothesis, she collected data from a sample of 37 adults who participated in an alcohol moderation management program. Alcohol consumption was measured as number of alcoholic drinks per week. (A standard definition is that one alcoholic drink contains 14 grams of alcohol, the equivalent of 12 ounces of regular beer or five ounces of wine.) Because younger and more educated people usually consume more alcohol, she controlled for age and education, both measured in years. The hypothetical data are shown in the following table:

Subject ID	Number of Drinks	Employment Status	Age	Education
1	2	Employed	46	12
2	3	Employed	50	16
3	0	Employed	48	12
4	7	Unemployed	23	16
5	3	Employed	30	14
6	5	Employed	34	18
7	2	Not in labor force	44	11
8	1	Employed	42	10
9	0	Employed	45	12
10	0	Not in labor force	42	12
11	8	Unemployed	22	16
12	12	Not in labor force	24	13
13	12	Unemployed	29	21
14	3	Not in labor force	30	20
15	4	Unemployed	32	16
16	5	Employed	32	22
17	10	Unemployed	21	12
18	13	Unemployed	27	20
19	9	Not in labor force	28	16
20	7	Not in labor force	30	18

(Continued)

Subject ID	Number of Drinks	Employment Status	Age	Education
21	5	Unemployed	40	16
22	5	Employed	30	14
23	5	Employed	29	14
24	3	Employed	45	12
25	2	Unemployed	46	12
26	5	Not in labor force	29	20
27	7	Employed	29	16
28	7	Not in labor force	28	18
29	0	Not in labor force	51	13
30	0	Employed	40	12
31	1	Employed	50	18
32	14	Unemployed	28	22
33	5	Not in labor force	60	20
34	2	Employed	24	16
35	3	Employed	22	12
36	4	Not in labor force	65	13
37	12	Unemployed	38	20

1. Analyze the data using multiple regression in a way that accounts for the dependent variable number of drinks per week being a count.
2. Interpret the results of the analysis in Question 1 as fully as possible.

8.4.2 Exercise 2

1. Suppose that X is a dummy independent variable for "employed" (with reference "unemployed"), and the associated b was statistically significant. Give the proper interpretation of the following results in a Poisson regression model for number of traffic citations in some time period:

   (a) b = 0.19

   (b) b = −0.15.

2. Suppose that X is a numerical independent variable for annual income (in $1,000s) and the associated b was statistically significant. Give the

proper interpretation of the following results in a Poisson model for number of traffic citations in some time period:

(a) b = 0.008

(b) b = −0.021.

8.4.3 Exercise 3: Transportation Use Patterns

A researcher was studying transportation use patterns, focusing on the annual number of trips that residents of a city made via train to the nearest larger city. He analyzed data from a sample of 70 residents of the city, predicting the number of train trips taken in the previous year from the independent variables gender (female or male), age, number of commercial plane rides taken that year, and car ownership (yes, shared with another household member, and no). Results of a Poisson regression analysis are as follows:

Chi-square = 91.25
df = 64
Chi-square/df = 1.426
Deviance = 108.09

	b	SE of b	Wald	p-value
Intercept (constant)	0.769	0.199		
Male	0.067	0.160	0.178	0.673
Age	0.025	0.172	0.021	0.885
Plane rides	0.163	0.060	7.361	0.007
Owns car	−0.692	0.178	15.095	0.000 (< 0.001)
Shared car	0.140	0.206	0.464	0.496

For a model without the car ownership dummies, deviance = 128.23.

1. Interpret results as fully as possible.

2. Based on these results, what would be sensible additional analyses for the researcher to perform?

9

A Brief Tour of Some Related Methods

In this chapter we briefly introduce a variety of methods and techniques that are related to or further extend the multiple regression framework. This list is certainly not comprehensive, and of course these introductions lack many details. Still, we hope that these descriptions will help acquaint readers with types of analyses that they otherwise might not be aware of and encourage them to learn more about these techniques in more specialized or advanced texts. Readers who recognize that a research question of great interest to them could be addressed with a particular method will be especially motivated to learn more. But even if that does not happen, we hope that some awareness of these methods will help leave readers from various fields better able to read and appreciate the published literature in their disciplines.

9.1 Regression Diagnostics

In the previous chapters we sometimes briefly mentioned some *regression diagnostics* that can in certain instances help us refine our analyses. When discussing logs in Chapter 5, we noted that examination of the residuals from a regression model may sometimes help us decide whether to log some variable(s), and in the Appendix to Chapter 2 we introduced a diagnostic measure that is often used to assess the possible impact of collinearity. These are just two examples from a much wider array of diagnostic tools that can be used in conjunction with multiple regression. Many of these diagnostic tools are available in the statistical software packages commonly used by social scientists.

One area in which diagnostics can be helpful is in the identification of outliers in our data set. Outliers are cases that, when graphed in a scatterplot, look distinctive relative to the rest of the cloud of points. They might have an unusually high or low value on the dependent variable, or on one or more independent variables, or both. The concern with outliers is the possibility that regression results may be unduly influenced by the inclusion of a single case, or small number of cases, in our sample. If so, we might question the generalizability of the findings, because they seem to stem substantially from the presence of this single case (or these few cases) in a way that seems unlikely to be repeated in a new sample. The presence of influential cases like this is probably a more realistic possibility in small data sets, such as a sample of 35 countries, than in large-scale surveys that include thousands of respondents. In a large data set, it is much less likely that the presence or absence of a single case could make much difference in our regression results. The notion of influential cases can be extended from this image of simple (bivariate) regression and the scatterplot to multiple regression.

Measures such as Cook's D and DFFITS are used to identify influential cases and can be chosen as options with multiple regression analyses in standard software. Texts that focus on such measures give rule-of-thumb cutoffs for determining what values of these measures are needed to flag a case as influential; cutoffs may depend on both the sample size and the number of independent variables in the regression. Other measures focus more specifically on whether a case is unusual in terms of its values on the independent variables, as such a case at least has the potential to affect regression results; this potential is referred to as leverage. Measures such as the studentized residual indicate whether a case's value on the dependent variable is markedly "discrepant" from the pattern established by the other N – 1 cases in the data. Note that the names of these measures are not completely uniform across software packages, so it is important to read software documentation carefully.

If a case is identified as influential, the researcher still would be reluctant to remove it from the analysis, especially if there were already little data available to analyze. But it may be helpful to report how results changed with and without this case, so that readers of the research article or report can include that in their understanding of the research findings, and it would be worth rechecking data sources to be sure that the data for that case were recorded accurately. If the study is using aggregate-level data, like from cities or countries, then it may be possible to use background knowledge of these aggregate units to think further about a discrepant case, perhaps finding additional independent variable(s) that account for the discrepancy. Identification of outliers should be a step toward better understanding the data rather than a trigger to throw out cases. As mentioned above, there are many other diagnostic tools with different goals that can help give other kinds of insights into our sample data.

9.2 Time-Series Analysis

The data in the examples and exercises throughout the book are *cross-sectional* in nature. That is, the measurements of the variables were taken at the same time (or as close as practical to the same time) for all members of the sample. The data were therefore a snapshot of the N units in the sample at essentially a single point in time. For example, in a city-level study of unemployment and crime in the United States, we would probably have measures of unemployment, crime, and our other variables taken in the same year for all of the cities in our sample.

An alternative type of data would report on measurements of variables over time for a single unit. In the unemployment and crime example, this could mean collecting annual measures of unemployment and crime for the United States as a whole over the past 65 years. That is, the unit of analysis would be the entire United States, with 65 annual observations on this one unit, rather than the units being the 50 states or some set of American cities. We could imagine there being for each variable a jagged graph showing its value across all the years in our data, with the jaggedness of the graph reflecting the pattern of year-to-year changes in the variable. Such data are often referred to as *time-series* data.

At first glance, it appears that we could simply use time-series data in a regression analysis just like we used cross-sectional data. One series would be used as the dependent variable, and other series would be the independent variables; in terms of the graphs of these variables over time, we would be looking at how one graph is predicted from the others. However, the results of this sort of naïve application of regression (or, even more basically, correlation) analysis to time-series data are likely to be of questionable validity. This concern stems from the very nature of time-series data.

For a series like the annual United States violent crime rate, we expect the values to typically change relatively little from one year to the next. There certainly are year-to-year fluctuations, and there is substantial media attention and public interest surrounding each year's announcement of whether official crime statistics went up or down from the previous year. But relative to the total variation in crime rates observed over the entire scope of many years of annual crime data, rates in one year will tend to be rather similar to those in the previous or next. That is, we will probably not see one of the highest-ever crime years being immediately followed by one of the lowest ever, and vice versa; the graph will be somewhat jagged, but not *that* jagged.

Of course more extreme variation could happen with, say, the annual homicide rate in a small community. Then a year with three homicides, implying quite a high homicide rate for a small population, could easily be followed immediately by a year with zero. And even in data representing a broader geographic area, large changes from one period to the next are not impossible, with some data contexts much more prone to these than are national crime rates. The 1929 stock market crash, for example, resulted in a very large and quick drop in stock index values and would be readily visible in a year-to-year graph. But for many social phenomena, nearby time periods are likely to show similar values.

One consequence of this year-to-year similarity is that for a regression using the annual crime rate as Y and the annual unemployment rate and other indicators as X's, the errors or residuals from neighboring years are likely to be similar too. If our regression predicts a lower crime rate for 1985 than what actually occurred, it is likely to underpredict crime for 1986 as well. Of course this will not necessarily be the case for all pairs of adjacent years in our data, but in general these errors for neighboring years will tend to be substantially correlated. This kind of *"autocorrelation"* (with the residuals in effect correlated with themselves) is symptomatic of a violation of multiple regression's underlying technical assumptions. Autocorrelation can be apparent in other contexts too, such as "spatial autocorrelation" in a study of city neighborhood data, where errors for close neighborhoods tend to be similar. But it is especially prevalent in analyses of time-series data.

There are many methods that try to correct for this issue in regression analyses of time-series data. Some are relatively straightforward, though these simpler approaches will not be appropriate in all circumstances. For instance, one approach is to analyze year-to-year differences in the dependent variable, rather than the dependent variable itself. That is, rather than taking Y_{1970} as the record for 1970, the analysis would use $(Y_{1970} - Y_{1969})$ instead, with similar differencing used for the independent variables as well. Other analyses might include the previous year's Y (the "lagged" value) among the independent variables predicting Y. There are also a variety of very complex solutions to this issue, and many advanced textbooks are devoted to analysis of time-series data, addressing autocorrelation and other technical issues that arise in such analysis. Today's standard statistical packages in social science have considerable capabilities for such analysis, but be sure to understand the theoretical underpinnings of any method of time-series analysis before applying it to data.

9.3 Multilevel Models

Our discussion of levels of analysis focused on the distinction between individual and aggregate levels, emphasizing that results obtained from data at one level cannot be assumed to apply to the other level too. There are instances, however, in which the predictors of some individual-level outcome include aggregate-level factors. The classic example often used to introduce this idea is that of predicting achievement (test scores, for instance) for individual school students. Surely there are a variety of individual-level factors that likely affect a student's performance on an achievement test, perhaps including elements of the student's home environment, physical attributes, demographic characteristics, or past academic assessments. But there may also be factors that operate above the level of individual students yet still affect their performance; these could include various social, demographic, and economic characteristics of the student's community, or organizational characteristics of the student's school. A comprehensive model for student performance would therefore be *multilevel*, with independent variables that focus on individual students as well as independent variables that reflect aggregate-level characteristics of the schools and communities to which the students belong.

This classic example illustrates the situation in which the individual-level units in our study are clustered within aggregate-level units, with the individual-level Y potentially related to independent variables at each level. One might think that we can simply include the aggregate-level variables as if they were additional individual-level variables and proceed with our regression as usual.

This would be reasonable if in the example our data came from a national random sample in which we would be very unlikely to have sampled two students from the same school. But a more typical situation would be for our data to have many students from each of a sample of schools, and in that case simply ignoring the difference in levels is almost certain to violate some of the underlying technical assumptions of multiple regression. That is especially threatening to our standard errors and hypothesis (significance) tests.

Multilevel models (sometimes called hierarchical models) address these concerns and permit estimation of the effects of both levels' variables. There is additional flexibility because it is also possible to create an aggregate-level variable from the mean of an individual-level variable in each aggregate-level unit. For instance, official data on a school will probably not include a measure of the mean education of its students' parents, but a school-level variable of that sort can be constructed from the responses on parents' education that the school's students gave in their surveys. Further, that individual-level variable can then be centered separately in each aggregate-level unit; for example, it could report how much higher or lower the education of a student's parents is than the mean parents' education in that school. With that construction of an aggregate-level variable along with centering of the individual-level variable, the analysis can distinguish within-group from between-group effects of that individual-level variable (where "group" refers to the aggregate-level unit). Multilevel models can also accommodate cross-level interaction, in which the effect of an individual-level variable on Y varies across the aggregate-level units as a function of some of the aggregate-level variables.

Multilevel models can be set up for different types of dependent variables, so it is possible to have, for instance, multilevel logistic regression along with the standard multilevel regression model for a continuous dependent variable. Texts on this topic cover the many details that must be addressed in implementing multilevel models, and this approach can be carried out in a variety of software packages (including some that are built especially for this purpose). Different software packages sometimes use slightly different terminology with these models, so it is important to take care in translating ideas from textbooks into software commands.

There are a variety of other data contexts that are similar in some respects to the multilevel situation. For example, perhaps the individual-level units are clustered, but the analyst simply wants the individual-level regression results to respect this fact, rather than actually using any aggregate-level predictors. Again the main goal in accounting for the clustering is to get the correct standard errors, with ongoing technical debates about how to most effectively do that. Longitudinal data, in which a series of measurements is taken on each individual unit, parallel multilevel data, but with a twist: the individual units now play what was the role of the aggregate-level units, and the different measurement occasions for an individual are in what had been the role of the individual-level units. The need to account for time, however, can make the longitudinal data situation somewhat more complicated than the traditional multilevel setting. (Notice the difference from the situation discussed above for time-series data, in which data consist of series of observations on just one unit.)

9.4 Survival Analysis

Survival analysis, or *event history analysis*, is used when the dependent variable is a measurement of time until some event. In some applications, survival is literally being observed, as in a medical study that tracks patients' time to death or an engineering study investigating time until failure of a machine part. But the method is also used in settings that do not literally involve survival, such as research on the length of time that a person remains unemployed after losing their job. "Event history analysis" is therefore also used among social scientists as a more encompassing term than "survival analysis."

At first glance this situation might not seem to require any special methods. In the context of time until finding a new job, one can imagine setting up a regression analysis in which the dependent variable is the length of the spell of unemployment, and the independent variables are various characteristics of the person that could predict how long it took them to find a new job. (Perhaps the dependent variable—time to a new job—would be logged in this scenario.) But this situation is actually more complicated, because, in most actual studies at least, data collection needs to be ended at some point. That is, the researchers may have decided to track the individuals for just one year; this could be due to financial constraints on the research or simply a desire not to have to wait too long to produce papers or reports describing the study. In that case, some individuals may not have found a new job by the time the study ends. Similarly, some patients may still be alive when the medical study concludes, or a machine part may still be working when the engineers stop observing it.

This phenomenon is referred to as censoring, so that an actual value of time until the event is not recorded for some individuals in the study, only an indication that the event had not yet occurred when the study's data collection ended. If we were simply carrying out a standard regression using the times (or logged times) to the event as Y, we would not know how to deal with such cases. A main appeal of survival analysis methods is therefore to provide a formal way of accommodating the existence of censoring when analyzing data on time to some event. The formal methods are constructed to estimate the hazard rate for an individual case at a particular time, where the hazard rate is, if we may be somewhat loose with the technical definition, akin to an instantaneous probability of the event occurring. A lower hazard rate implies a longer expected time to the event.

Typical models are set up as regressions for the log hazard rate, so that e^b is interpreted as the multiplicative effect of X on the hazard rate. (This "hazard ratio" is analogous to the odds ratio in logistic regression.) There are many variations on this framework, accommodating different kinds of data and different hypothesized processes. Some variations include modifications for discrete vs. continuous measurement of time, whether or not the event of interest can, by definition, occur only once to an individual or is possibly recurring, and whether or not the model specifies the form of time dependence in the process (such as the event becoming increasingly or decreasingly likely to occur as time passes) or leaves the nature of the time dependence in the process unspecified. Survival analysis procedures are available in all of the statistical software packages typically used by social scientists, though not all packages support every specific technique.

9.5 Principal Components

Suppose that the survey data used in an analysis include several clearly related questions. For instance, survey respondents are sometimes asked a series of questions exploring their feelings about the death penalty, with different questions asking about support for the death penalty in the context of various characteristics of crimes or victims. Although the different questions may be exploring important nuance in respondents' beliefs about this issue, we also expect that in general the responses are likely to be highly correlated. That is, people who support or oppose the death penalty in one set of circumstances are likely to feel the same way for different circumstances. In that case, the analyst might decide to turn these questions into a single scale by adding up the responses given for each (reversing the responses where necessary, if saying "agree" sometimes meant support for the death penalty and sometimes opposition). If the responses to the different questions are highly correlated, this probably will not risk too much loss of information, and it will be possible for analyses to include a single variable summarizing respondents' feelings about this issue, instead of a long list of a variables focusing on specific contexts.

Principal components are a way to formalize this idea of combining information from several variables. In principal component analysis, a set of variables is re-expressed as a set of new variables called principal components. The set of principal components together represents the same total variance that was present in the original set of variables, and each principal component is a weighted sum of the original set of variables. That is, a given principal component may reflect some of the original variables much more than others, unlike the simple scale above in which each original variable implicitly had the same weight when making the new variable. This will help us interpret the meaning of each of these new variables, as we can see which components are most highly correlated with which original variables.

The method of constructing the principal components results in the first—most important—component representing as much of the total variance in the original set of variables as possible. The second component is constructed to have zero correlation (in the sample) with the first component while accounting for as much of the remaining original variance as possible. This process continues, until as many components have been constructed as there were original variables, and all of the original variables' variance is accounted for by the components.

Each component is uncorrelated (in the sample) with any other, and the components represent descending shares of the total variance in the set of original variables. (In practice, the set of original variables may include very different kinds of measurements, so that their variances are not really comparable. In that case, the original variables would be standardized to have equal variances before performing the principal component analysis.) In some cases, a small number of components will account for a very large proportion of the original variables' variance, which will be helpful for interpreting the meaning of the components.

The fact that the components are constructed so as to be uncorrelated with each other in the sample means that this can be one response to collinearity in multiple regression. If we have a set of highly correlated X's, we can construct principal components from these X's, and the resulting components are by definition exactly uncorrelated with each other. If we use these components as independent variables, instead of the original X's, we know that collinearity cannot be a problem within this new set. (The components could still be correlated with other X's that remain in the regression too, but we could of course include these other X's in the construction of the components if desired.) And if we feel that we can use just a small number of the components without losing much information, the regression will also be less complicated in the sense of having fewer X's.

This sounds very good, but there is a cost as well. When we include principal components as independent variables in our regression, it may be hard to interpret the components' effects on the predicted Y. The components will not have the usual units of X that we normally use in our interpretation, and sometimes we do not have a very clear understanding of exactly how to make sense of a particular component, which in turn makes it difficult to understand its effect on Y. We might think we could take the formula relating a component to the original X's and plug that into the regression equation, in order to bring the original X's back into view in the regression equation. But in general that will not be an appropriate way to determine the effects of the original X's. So the tradeoff is that use of the components as independent variables can eliminate the collinearity problem, but they can also make it much harder to interpret regression results.

Principal component analysis is available in any of the standard statistical software packages; it is important to study the software documentation to be sure that one is producing the intended analysis. We also note that principal component analysis shares much the same spirit as another technique known as factor analysis. The main difference is that principal component analysis takes the perspective of combining the observed variables in our data to form new variables, but factor analysis takes the perspective that there are unobserved (latent) factors that are reflected (imperfectly) in the observed variables. The construction of principal components does not rely on any idea of latent variables, while this concept is central to the theory of factor analysis. Still, despite this underlying difference, in many data sets the stories that principal component analysis and factor analysis tell seem quite similar, and there are also technical connections between the two methods.

9.6 Structural Equations

There is a long tradition in the social sciences of "path diagrams" representing a series of regression models, in which the independent variables for a regression predicting the ultimate outcome of interest take the role of dependent variable in other regressions. For example, suppose that the final outcome in such a system is income, and that education and mother's education are among the independent variables predicting it. Suppose that in another regression, education is the dependent variable and mother's education is an independent variable. Then the path diagram would show a direct effect of mother's education on income (from the first regression), but also an indirect effect, via its direct effect on education and then in turn education's direct effect on income. Theoretical ideas about which of these variables "cause" other variables can be represented in such a path diagram.

Structural equation models update this traditional approach in a variety of ways and address technical problems that can result when simply running a series of conventional regression models in order to create a path model. One benefit is that variables in this sort of system can be conceived of as including disturbance, or error, terms, and different patterns of correlations among the variables' disturbances can be hypothesized by the researcher. Another is that there is more flexibility in specifying how two variables are related, including a causal effect of one on the other (often diagrammed as a single-headed arrow from one variable to another), a correlation without a causal direction specified (a double-headed arrow), or as mutual causation, or direct feedback, in which each can be said to have a causal effect on the other (represented by two arrows pointing in opposite directions).

An especially important aspect of the modern framework is that these structural relations can be hypothesized for latent (unobserved) variables. That is, the researcher's theory of what affects what can include relations among latent variables, rather than only concerning observed variables as in traditional path analysis. Although a latent variable is, by definition, not directly observed, each hypothesized latent variable is assumed to have observed indicators, which are believed by the researcher to be manifestations of the underlying latent variable. For example, a researcher may hypothesize that a subject's response to a survey question on willingness to use force in a confrontation is one indicator of a latent variable "aggressiveness."

This linking of latent variables to observed indicators is similar to that in factor analysis (mentioned above in the section on principal components). In the structural equations framework, this is referred to as "confirmatory factor analysis," in which the researcher has hypothesized about which particular observed variables are related to which latent factors. The factor analysis mentioned earlier is, in contrast, referred to as "exploratory factor analysis," in which potentially any of the observed variables could be related to any of the factors, in a pattern that is not specified by the researcher in advance. Structural equation models can thus link a structural model for regression-like relationships among latent factors with a confirmatory factor analysis model that relates observed indicators to the latent factors.

This framework can get very complicated, with many important technical issues that must be kept in mind when using it in research. But this approach is also very general in its application, so that the framework conceptually incorporates many kinds of seemingly distinct analytic techniques. Among the challenging aspects is that a hypothesized model may be completely reasonable in theoretical terms, yet still cannot be estimated from actual data. Setting up models and interpreting their results is less straightforward than in other types of analysis that we have discussed, and a researcher probably needs to study this approach quite a bit before becoming comfortable applying it. Nonetheless, the possibilities for representing theory and estimating the implied models with data make this an appealing approach to many social scientists. Note that standard software packages have differing capabilities for such analyses, and specialized software is also popular in this area, so particular care should be taken in perusing software documentation for this topic.

9.7 Causal Inference

Throughout the book, we have noted that the language of an "effect" of X on Y in multiple regression should be taken as a convenient usage, or as a statement that concerns our predictions in the regression model, rather than implying that the analysis has uncovered a genuine causal effect of X on Y. That is because most social science applications involve observational rather experimental data, and cases in the sample were not randomly assigned a level of X as they would be in a true experiment. Even with a comprehensive set of theoretically well-justified control variables, we are reluctant to assert that a regression analysis has demonstrated a legitimate causal effect of X on Y.

In recent decades, social scientists have grown interested in methods that hope to estimate causal effects in analyses of nonexperimental data. One area of this interest is in so-called "natural experiments," in which the execution of a policy or some other circumstances outside of researchers' control produce a research design in which assignment of cases to different levels of a treatment variable seems to mimic random assignment in a true experiment. Another area of interest is in the extension of matching methods, in which cases that did receive some treatment are matched to cases that did not receive the treatment but are otherwise similar. It is then hoped that the causal effect of treatment can be estimated from the difference in average outcomes between the matched groups. Simple matching of treated and untreated cases on a small number of observed characteristics is not likely to be very effective, especially because true random assignment will produce treated and untreated groups that are, on average, matched in both measured and unmeasured characteristics. But more sophisticated strategies such as "propensity score matching" attempt to better emulate this by matching treated and untreated cases that had the same estimated propensity to seek treatment. The goal of these methods is to attack the selection problem in which those who seek treatment differ from those who do not, or differ in their expected benefits from treatment.

In something like the more familiar regression context, there has been considerable attention to how hypothesized relationships between, first, the independent variable of main interest and control variables, and, second, control variables and the main outcome variable do or do not allow the researcher to identify a causal effect of the main independent variable on the outcome. This can involve diagrams similar to those in path analysis and structural equations, and this approach is linked to the analytic framework of structural equations.

Note that these techniques involve investigation of hypothesized causal structures; they do not automatically reveal causal patterns to the researcher.

There is ongoing development of the various methods sketched here, and the whole enterprise of asserting causal effects from nonexperimental data that were not produced via classical random assignment remains controversial. Not all researchers find the results of this sort of analysis compelling, so research design remains a key element. But nonetheless it is clear that there is widening interest in these approaches among social scientists today. Advanced texts in this area provide background and many details on these methods.

APPENDIX A ● F Distribution

denominator degrees of freedom	p	numerator degrees of freedom 1	2	3	4	5	6	7	8
1	.100	39.86	49.50	53.59	55.83	57.24	58.20	58.91	59.44
	.050	161.45	199.50	215.71	224.58	230.16	233.99	236.77	238.88
	.025	647.79	799.50	864.16	899.58	921.85	937.11	948.22	956.66
	.010	4052.2	4999.5	5403.4	5624.6	5763.6	5859.0	5928.4	5981.1
	.001	405284	500000	540379	562500	576405	585937	592873	598144
2	.100	8.53	9.00	9.16	9.24	9.29	9.33	9.35	9.37
	.050	18.51	19.00	19.16	19.25	19.30	19.33	19.35	19.37
	.025	38.51	39.00	39.17	39.25	39.30	39.33	39.36	39.37
	.010	98.50	99.00	99.17	99.25	99.30	99.33	99.36	99.37
	.001	998.50	999.00	999.17	999.25	999.30	999.33	999.36	999.37
3	.100	5.54	5.46	5.39	5.34	5.31	5.28	5.27	5.25
	.050	10.13	9.55	9.28	9.12	9.01	8.94	8.89	8.85
	.025	17.44	16.04	15.44	15.10	14.88	14.73	14.62	14.54
	.010	34.12	30.82	29.46	28.71	28.24	27.91	27.67	27.49
	.001	167.03	148.50	141.11	137.10	134.58	132.85	131.58	130.62
4	.100	4.54	4.32	4.19	4.11	4.05	4.01	3.98	3.95
	.050	7.71	6.94	6.59	6.39	6.26	6.16	6.09	6.04
	.025	12.22	10.65	9.98	9.60	9.36	9.20	9.07	8.98
	.010	21.20	18.00	16.69	15.98	15.52	15.21	14.98	14.80
	.001	74.14	61.25	56.18	53.44	51.71	50.53	49.66	49.00
5	.100	4.06	3.78	3.62	3.52	3.45	3.40	3.37	3.34
	.050	6.61	5.79	5.41	5.19	5.05	4.95	4.88	4.82
	.025	10.01	8.43	7.76	7.39	7.15	6.98	6.85	6.76
	.010	16.26	13.27	12.06	11.39	10.97	10.67	10.46	10.29
	.001	47.18	37.12	33.20	31.09	29.75	28.83	28.16	27.65
6	.100	3.78	3.46	3.29	3.18	3.11	3.05	3.01	2.98
	.050	5.99	5.14	4.76	4.53	4.39	4.28	4.21	4.15
	.025	8.81	7.26	6.60	6.23	5.99	5.82	5.70	5.60
	.010	13.75	10.92	9.78	9.15	8.75	8.47	8.26	8.10
	.001	35.51	27.00	23.70	21.92	20.80	20.03	19.46	19.03
7	.100	3.59	3.26	3.07	2.96	2.88	2.83	2.78	2.75
	.050	5.59	4.74	4.35	4.12	3.97	3.87	3.79	3.73
	.025	8.07	6.54	5.89	5.52	5.29	5.12	4.99	4.90
	.010	12.25	9.55	8.45	7.85	7.46	7.19	6.99	6.84
	.001	29.25	21.69	18.77	17.20	16.21	15.52	15.02	14.63
8	.100	3.46	3.11	2.92	2.81	2.73	2.67	2.62	2.59
	.050	5.32	4.46	4.07	3.84	3.69	3.58	3.50	3.44
	.025	7.57	6.06	5.42	5.05	4.82	4.65	4.53	4.43
	.010	11.26	8.65	7.59	7.01	6.63	6.37	6.18	6.03
	.001	25.41	18.49	15.83	14.39	13.48	12.86	12.40	12.05

(Continued)

APPENDIX A ● F Distribution (*Continued*)

denominator degrees of freedom	p	numerator degrees of freedom 9	10	12	15	20	25	30	40	50
1	.100	59.86	60.19	60.71	61.22	61.74	62.05	62.26	62.53	62.69
	.050	240.54	241.88	243.91	245.95	248.01	249.26	250.10	251.14	251.77
	.025	963.28	968.63	976.71	984.87	993.10	998.08	1001.4	1005.6	1008.1
	.010	6022.5	6055.8	6106.3	6157.3	6208.7	6239.8	6260.6	6286.8	6302.5
	.001	602284	605621	610668	615764	620908	624017	626099	628712	630285
2	.100	9.38	9.39	9.41	9.42	9.44	9.45	9.46	9.47	9.47
	.050	19.38	19.40	19.41	19.43	19.45	19.46	19.46	19.47	19.48
	.025	39.39	39.40	39.41	39.43	39.45	39.46	39.46	39.47	39.48
	.010	99.39	99.40	99.42	99.43	99.45	99.46	99.47	99.47	99.48
	.001	999.39	999.40	999.42	999.43	999.45	999.46	999.47	999.47	999.48
3	.100	5.24	5.23	5.22	5.20	5.18	5.17	5.17	5.16	5.15
	.050	8.81	8.79	8.74	8.70	8.66	8.63	8.62	8.59	8.58
	.025	14.47	14.42	14.34	14.25	14.17	14.12	14.08	14.04	14.01
	.010	27.35	27.23	27.05	26.87	26.69	26.58	26.50	26.41	26.35
	.001	129.86	129.25	128.32	127.37	126.42	125.84	125.45	124.96	124.66
4	.100	3.94	3.92	3.90	3.87	3.84	3.83	3.82	3.80	3.80
	.050	6.00	5.96	5.91	5.86	5.80	5.77	5.75	5.72	5.70
	.025	8.90	8.84	8.75	8.66	8.56	8.50	8.46	8.41	8.38
	.010	14.66	14.55	14.37	14.20	14.02	13.91	13.84	13.75	13.69
	.001	48.47	48.05	47.41	46.76	46.10	45.70	45.43	45.09	44.88
5	.100	3.32	3.30	3.27	3.24	3.21	3.19	3.17	3.16	3.15
	.050	4.77	4.74	4.68	4.62	4.56	4.52	4.50	4.46	4.44
	.025	6.68	6.62	6.52	6.43	6.33	6.27	6.23	6.18	6.14
	.010	10.16	10.05	9.89	9.72	9.55	9.45	9.38	9.29	9.24
	.001	27.24	26.92	26.42	25.91	25.39	25.08	24.87	24.60	24.44
6	.100	2.96	2.94	2.90	2.87	2.84	2.81	2.80	2.78	2.77
	.050	4.10	4.06	4.00	3.94	3.87	3.83	3.81	3.77	3.75
	.025	5.52	5.46	5.37	5.27	5.17	5.11	5.07	5.01	4.98
	.010	7.98	7.87	7.72	7.56	7.40	7.30	7.23	7.14	7.09
	.001	18.69	18.41	17.99	17.56	17.12	16.85	16.67	16.44	16.31
7	.100	2.72	2.70	2.67	2.63	2.59	2.57	2.56	2.54	2.52
	.050	3.68	3.64	3.57	3.51	3.44	3.40	3.38	3.34	3.32
	.025	4.82	4.76	4.67	4.57	4.47	4.40	4.36	4.31	4.28
	.010	6.72	6.62	6.47	6.31	6.16	6.06	5.99	5.91	5.86
	.001	14.33	14.08	13.71	13.32	12.93	12.69	12.53	12.33	12.20
8	.100	2.56	2.54	2.50	2.46	2.42	2.40	2.38	2.36	2.35
	.050	3.39	3.35	3.28	3.22	3.15	3.11	3.08	3.04	3.02
	.025	4.36	4.30	4.20	4.10	4.00	3.94	3.89	3.84	3.81
	.010	5.91	5.81	5.67	5.52	5.36	5.26	5.20	5.12	5.07
	.001	11.77	11.54	11.19	10.84	10.48	10.26	10.11	9.92	9.80

		numerator degrees of freedom							
denominator degrees of freedom	p	1	2	3	4	5	6	7	8
9	.100	3.36	3.01	2.81	2.69	2.61	2.55	2.51	2.47
	.050	5.12	4.26	3.86	3.63	3.48	3.37	3.29	3.23
	.025	7.21	5.71	5.08	4.72	4.48	4.32	4.20	4.10
	.010	10.56	8.02	6.99	6.42	6.06	5.80	5.61	5.47
	.001	22.86	16.39	13.90	12.56	11.71	11.13	10.70	10.37
10	.100	3.29	2.92	2.73	2.61	2.52	2.46	2.41	2.38
	.050	4.96	4.10	3.71	3.48	3.33	3.22	3.14	3.07
	.025	6.94	5.46	4.83	4.47	4.24	4.07	3.95	3.85
	.010	10.04	7.56	6.55	5.99	5.64	5.39	5.20	5.06
	.001	21.04	14.91	12.55	11.28	10.48	9.93	9.52	9.20
11	.100	3.23	2.86	2.66	2.54	2.45	2.39	2.34	2.30
	.050	4.84	3.98	3.59	3.36	3.20	3.09	3.01	2.95
	.025	6.72	5.26	4.63	4.28	4.04	3.88	3.76	3.66
	.010	9.65	7.21	6.22	5.67	5.32	5.07	4.89	4.74
	.001	19.69	13.81	11.56	10.35	9.58	9.05	8.66	8.35
12	.100	3.18	2.81	2.61	2.48	2.39	2.33	2.28	2.24
	.050	4.75	3.89	3.49	3.26	3.11	3.00	2.91	2.85
	.025	6.55	5.10	4.47	4.12	3.89	3.73	3.61	3.51
	.010	9.33	6.93	5.95	5.41	5.06	4.82	4.64	4.50
	.001	18.64	12.97	10.80	9.63	8.89	8.38	8.00	7.71
13	.100	3.14	2.76	2.56	2.43	2.35	2.28	2.23	2.20
	.050	4.67	3.81	3.41	3.18	3.03	2.92	2.83	2.77
	.025	6.41	4.97	4.35	4.00	3.77	3.60	3.48	3.39
	.010	9.07	6.70	5.74	5.21	4.86	4.62	4.44	4.30
	.001	17.82	12.31	10.21	9.07	8.35	7.86	7.49	7.21
14	.100	3.10	2.73	2.52	2.39	2.31	2.24	2.19	2.15
	.050	4.60	3.74	3.34	3.11	2.96	2.85	2.76	2.70
	.025	6.30	4.86	4.24	3.89	3.66	3.50	3.38	3.29
	.010	8.86	6.51	5.56	5.04	4.69	4.46	4.28	4.14
	.001	17.14	11.78	9.73	8.62	7.92	7.44	7.08	6.80
15	.100	3.07	2.70	2.49	2.36	2.27	2.21	2.16	2.12
	.050	4.54	3.68	3.29	3.06	2.90	2.79	2.71	2.64
	.025	6.20	4.77	4.15	3.80	3.58	3.41	3.29	3.20
	.010	8.68	6.36	5.42	4.89	4.56	4.32	4.14	4.00
	.001	16.59	11.34	9.34	8.25	7.57	7.09	6.74	6.47
16	.100	3.05	2.67	2.46	2.33	2.24	2.18	2.13	2.09
	.050	4.49	3.63	3.24	3.01	2.85	2.74	2.66	2.59
	.025	6.12	4.69	4.08	3.73	3.50	3.34	3.22	3.12
	.010	8.53	6.23	5.29	4.77	4.44	4.20	4.03	3.89
	.001	16.12	10.97	9.01	7.94	7.27	6.80	6.46	6.19

(Continued)

APPENDIX A ● F Distribution (*Continued*)

		numerator degrees of freedom								
denominator degrees of freedom	p	9	10	12	15	20	25	30	40	50
9	.100	2.44	2.42	2.38	2.34	2.30	2.27	2.25	2.23	2.22
	.050	3.18	3.14	3.07	3.01	2.94	2.89	2.86	2.83	2.80
	.025	4.03	3.96	3.87	3.77	3.67	3.60	3.56	3.51	3.47
	.010	5.35	5.26	5.11	4.96	4.81	4.71	4.65	4.57	4.52
	.001	10.11	9.89	9.57	9.24	8.90	8.69	8.55	8.37	8.26
10	.100	2.35	2.32	2.28	2.24	2.20	2.17	2.16	2.13	2.12
	.050	3.02	2.98	2.91	2.85	2.77	2.73	2.70	2.66	2.64
	.025	3.78	3.72	3.62	3.52	3.42	3.35	3.31	3.26	3.22
	.010	4.94	4.85	4.71	4.56	4.41	4.31	4.25	4.17	4.12
	.001	8.96	8.75	8.45	8.13	7.80	7.60	7.47	7.30	7.19
11	.100	2.27	2.25	2.21	2.17	2.12	2.10	2.08	2.05	2.04
	.050	2.90	2.85	2.79	2.72	2.65	2.60	2.57	2.53	2.51
	.025	3.59	3.53	3.43	3.33	3.23	3.16	3.12	3.06	3.03
	.010	4.63	4.54	4.40	4.25	4.10	4.01	3.94	3.86	3.81
	.001	8.12	7.92	7.63	7.32	7.01	6.81	6.68	6.52	6.42
12	.100	2.21	2.19	2.15	2.10	2.06	2.03	2.01	1.99	1.97
	.050	2.80	2.75	2.69	2.62	2.54	2.50	2.47	2.43	2.40
	.025	3.44	3.37	3.28	3.18	3.07	3.01	2.96	2.91	2.87
	.010	4.39	4.30	4.16	4.01	3.86	3.76	3.70	3.62	3.57
	.001	7.48	7.29	7.00	6.71	6.40	6.22	6.09	5.93	5.83
13	.100	2.16	2.14	2.10	2.05	2.01	1.98	1.96	1.93	1.92
	.050	2.71	2.67	2.60	2.53	2.46	2.41	2.38	2.34	2.31
	.025	3.31	3.25	3.15	3.05	2.95	2.88	2.84	2.78	2.74
	.010	4.19	4.10	3.96	3.82	3.66	3.57	3.51	3.43	3.38
	.001	6.98	6.80	6.52	6.23	5.93	5.75	5.63	5.47	5.37
14	.100	2.12	2.10	2.05	2.01	1.96	1.93	1.91	1.89	1.87
	.050	2.65	2.60	2.53	2.46	2.39	2.34	2.31	2.27	2.24
	.025	3.21	3.15	3.05	2.95	2.84	2.78	2.73	2.67	2.64
	.010	4.03	3.94	3.80	3.66	3.51	3.41	3.35	3.27	3.22
	.001	6.58	6.40	6.13	5.85	5.56	5.38	5.25	5.10	5.00
15	.100	2.09	2.06	2.02	1.97	1.92	1.89	1.87	1.85	1.83
	.050	2.59	2.54	2.48	2.40	2.33	2.28	2.25	2.20	2.18
	.025	3.12	3.06	2.96	2.86	2.76	2.69	2.64	2.59	2.55
	.010	3.89	3.80	3.67	3.52	3.37	3.28	3.21	3.13	3.08
	.001	6.26	6.08	5.81	5.54	5.25	5.07	4.95	4.80	4.70
16	.100	2.06	2.03	1.99	1.94	1.89	1.86	1.84	1.81	1.79
	.050	2.54	2.49	2.42	2.35	2.28	2.23	2.19	2.15	2.12
	.025	3.05	2.99	2.89	2.79	2.68	2.61	2.57	2.51	2.47
	.010	3.78	3.69	3.55	3.41	3.26	3.16	3.10	3.02	2.97
	.001	5.98	5.81	5.55	5.27	4.99	4.82	4.70	4.54	4.45

		numerator degrees of freedom							
denominator degrees of freedom	p	1	2	3	4	5	6	7	8
17	.100	3.03	2.64	2.44	2.31	2.22	2.15	2.10	2.06
	.050	4.45	3.59	3.20	2.96	2.81	2.70	2.61	2.55
	.025	6.04	4.62	4.01	3.66	3.44	3.28	3.16	3.06
	.010	8.40	6.11	5.19	4.67	4.34	4.10	3.93	3.79
	.001	15.72	10.66	8.73	7.68	7.02	6.56	6.22	5.96
18	.100	3.01	2.62	2.42	2.29	2.20	2.13	2.08	2.04
	.050	4.41	3.55	3.16	2.93	2.77	2.66	2.58	2.51
	.025	5.98	4.56	3.95	3.61	3.38	3.22	3.10	3.01
	.010	8.29	6.01	5.09	4.58	4.25	4.01	3.84	3.71
	.001	15.38	10.39	8.49	7.46	6.81	6.35	6.02	9.20
19	.100	2.99	2.61	2.40	2.27	2.18	2.11	2.06	2.02
	.050	4.38	3.52	3.13	2.90	2.74	2.63	2.54	2.48
	.025	5.92	4.51	3.90	3.56	3.33	3.17	3.05	2.96
	.010	8.18	5.93	5.01	4.50	4.17	3.94	3.77	3.63
	.001	15.08	10.16	8.28	7.27	6.62	6.18	5.85	5.59
20	.100	2.97	2.59	2.38	2.25	2.16	2.09	2.04	2.00
	.050	4.35	3.49	3.10	2.87	2.71	2.60	2.51	2.45
	.025	5.87	4.46	3.86	3.51	3.29	3.13	3.01	2.91
	.010	8.10	5.85	4.94	4.43	4.10	3.87	3.70	3.56
	.001	14.82	9.95	8.10	7.10	6.46	6.02	5.69	5.44
21	.100	2.96	2.57	2.36	2.23	2.14	2.08	2.02	1.98
	.050	4.32	3.47	3.07	2.84	2.68	2.57	2.49	2.42
	.025	5.83	4.42	3.82	3.48	3.25	3.09	2.97	2.87
	.010	8.02	5.78	4.87	4.37	4.04	3.81	3.64	3.51
	.001	14.59	9.77	7.94	6.95	6.32	5.88	5.56	5.31
22	.100	2.95	2.56	2.35	2.22	2.13	2.06	2.01	1.97
	.050	4.30	3.44	3.05	2.82	2.66	2.55	2.46	2.40
	.025	5.79	4.38	3.78	3.44	3.22	3.05	2.93	2.84
	.010	7.95	5.72	4.82	4.31	3.99	3.76	3.59	3.45
	.001	14.38	9.61	7.80	6.81	6.19	5.76	5.44	5.19
23	.100	2.94	2.55	2.34	2.21	2.11	2.05	1.99	1.95
	.050	4.28	3.42	3.03	2.80	2.64	2.53	2.44	2.37
	.025	5.75	4.35	3.75	3.41	3.18	3.02	2.90	2.81
	.010	7.88	5.66	4.76	4.26	3.94	3.71	3.54	3.41
	.001	14.20	9.47	7.67	6.70	6.08	5.65	5.33	5.09
24	.100	2.93	2.54	2.33	2.19	2.10	2.04	1.98	1.94
	.050	4.26	3.40	3.01	2.78	2.62	2.51	2.42	2.36
	.025	5.72	4.32	3.72	3.38	3.15	2.99	2.87	2.78
	.010	7.82	5.61	4.72	4.22	3.90	3.67	3.50	3.36
	.001	14.03	9.34	7.55	6.59	5.98	5.55	5.23	4.99

(*Continued*)

APPENDIX A ● F Distribution (*Continued*)

denominator degrees of freedom	p	9	10	12	15	20	25	30	40	50
		numerator degrees of freedom								
17	.100	2.03	2.00	1.96	1.91	1.86	1.83	1.81	1.78	1.76
	.050	2.49	2.45	2.38	2.31	2.23	2.18	2.15	2.10	2.08
	.025	2.98	2.92	2.82	2.72	2.62	2.55	2.50	2.44	2.41
	.010	3.68	3.59	3.46	3.31	3.16	3.07	3.00	2.92	2.87
	.001	5.75	5.58	5.32	5.05	4.78	4.60	4.48	4.33	4.24
18	.100	2.00	1.98	1.93	1.89	1.84	1.80	1.78	1.75	1.74
	.050	2.46	2.41	2.34	2.27	2.19	2.14	2.11	2.06	2.04
	.025	2.93	2.87	2.77	2.67	2.56	2.49	2.44	2.38	2.35
	.010	3.60	3.51	3.37	3.23	3.08	2.98	2.92	2.84	2.78
	.001	5.56	5.39	5.13	4.87	4.59	4.42	4.30	4.15	4.06
19	.100	1.98	1.96	1.91	1.86	1.81	1.78	1.76	1.73	1.71
	.050	2.42	2.38	2.31	2.23	2.16	2.11	2.07	2.03	2.00
	.025	2.88	2.82	2.72	2.62	2.51	2.44	2.39	2.33	2.30
	.010	3.52	3.43	3.30	3.15	3.00	2.91	2.84	2.76	2.71
	.001	5.39	5.22	4.97	4.70	4.43	4.26	4.14	3.99	3.90
20	.100	1.96	1.94	1.89	1.84	1.79	1.76	1.74	1.71	1.69
	.050	2.39	2.35	2.28	2.20	2.12	2.07	2.04	1.99	1.97
	.025	2.84	2.77	2.68	2.57	2.46	2.40	2.35	2.29	2.25
	.010	3.46	3.37	3.23	3.09	2.94	2.84	2.78	2.69	2.64
	.001	5.24	5.08	4.82	4.56	4.29	4.12	4.00	3.86	3.77
21	.100	1.95	1.92	1.87	1.83	1.78	1.74	1.72	1.69	1.67
	.050	2.37	2.32	2.25	2.18	2.10	2.05	2.01	1.96	1.94
	.025	2.80	2.73	2.64	2.53	2.42	2.36	2.31	2.25	2.21
	.010	3.40	3.31	3.17	3.03	2.88	2.79	2.72	2.64	2.58
	.001	5.11	4.95	4.70	4.44	4.17	4.00	3.88	3.74	3.64
22	.100	1.93	1.90	1.86	1.81	1.76	1.73	1.70	1.67	1.65
	.050	2.34	2.30	2.23	2.15	2.07	2.02	1.98	1.94	1.91
	.025	2.76	2.70	2.60	2.50	2.39	2.32	2.27	2.21	2.17
	.010	3.35	3.26	3.12	2.98	2.83	2.73	2.67	2.58	2.53
	.001	4.99	4.83	4.58	4.33	4.06	3.89	3.78	3.63	3.54
23	.100	1.92	1.89	1.84	1.80	1.74	1.71	1.69	1.66	1.64
	.050	2.32	2.27	2.20	2.13	2.05	2.00	1.96	1.91	1.88
	.025	2.73	2.67	2.57	2.47	2.36	2.29	2.24	2.18	2.14
	.010	3.30	3.21	3.07	2.93	2.78	2.69	2.62	2.54	2.48
	.001	4.89	4.73	4.48	4.23	3.96	3.79	3.68	3.53	3.44
24	.100	1.91	1.88	1.83	1.78	1.73	1.70	1.67	1.64	1.62
	.050	2.30	2.25	2.18	2.11	2.03	1.97	1.94	1.89	1.86
	.025	2.70	2.64	2.54	2.44	2.33	2.26	2.21	2.15	2.11
	.010	3.26	3.17	3.03	2.89	2.74	2.64	2.58	2.49	2.44
	.001	4.80	4.64	4.39	4.14	3.87	3.71	3.59	3.45	3.36

denominator degrees of freedom	p	1	2	3	4	5	6	7	8
25	.100	2.92	2.53	2.32	2.18	2.09	2.02	1.97	1.93
	.050	4.24	3.39	2.99	2.76	2.60	2.49	2.40	2.34
	.025	5.69	4.29	3.69	3.35	3.13	2.97	2.85	2.75
	.010	7.77	5.57	4.68	4.18	3.85	3.63	3.46	3.32
	.001	13.88	9.22	7.45	6.49	5.89	5.46	5.15	4.91
26	.100	2.91	2.52	2.31	2.17	2.08	2.01	1.96	1.92
	.050	4.23	3.37	2.98	2.74	2.59	2.47	2.39	2.32
	.025	5.66	4.27	3.67	3.33	3.10	2.94	2.82	2.73
	.010	7.72	5.53	4.64	4.14	3.82	3.59	3.42	3.29
	.001	13.74	9.12	7.36	6.41	5.80	5.38	5.07	4.83
27	.100	2.90	2.51	2.30	2.17	2.07	2.00	1.95	1.91
	.050	4.21	3.35	2.96	2.73	2.57	2.46	2.37	2.31
	.025	5.63	4.24	3.65	3.31	3.08	2.92	2.80	2.71
	.010	7.68	5.49	4.60	4.11	3.78	3.56	3.39	3.26
	.001	13.61	9.02	7.27	6.33	5.73	5.31	5.00	4.76
28	.100	2.89	2.50	2.29	2.16	2.06	2.00	1.94	1.90
	.050	4.20	3.34	2.95	2.71	2.56	2.45	2.36	2.29
	.025	5.61	4.22	3.63	3.29	3.06	2.90	2.78	2.69
	.010	7.64	5.45	4.57	4.07	3.75	3.53	3.36	3.23
	.001	13.50	8.93	7.19	6.25	5.66	5.24	4.93	4.69
29	.100	2.89	2.50	2.28	2.15	2.06	1.99	1.93	1.89
	.050	4.18	3.33	2.93	2.70	2.55	2.43	2.35	2.28
	.025	5.59	4.20	3.61	3.27	3.04	2.88	2.76	2.97
	.010	7.60	5.42	4.54	4.04	3.73	3.50	3.33	3.20
	.001	13.39	8.85	7.12	6.19	5.59	5.18	4.87	4.64
30	.100	2.88	2.49	2.28	2.14	2.05	1.98	1.93	1.88
	.050	4.17	3.32	2.92	2.69	2.53	2.42	2.33	2.27
	.025	5.57	4.18	3.59	3.25	3.03	2.87	2.75	2.65
	.010	7.56	5.39	4.51	4.02	3.70	3.47	3.30	3.71
	.001	13.29	8.77	7.05	6.12	5.53	5.12	4.82	4.58
40	.100	2.84	2.44	2.23	2.09	2.00	1.93	1.87	1.83
	.050	4.08	3.23	2.84	2.61	2.45	2.34	2.25	2.18
	.025	5.42	4.05	3.46	3.13	2.90	2.74	2.62	2.53
	.010	7.31	5.18	4.31	3.83	3.51	3.29	3.12	2.99
	.001	12.61	8.25	6.59	5.70	5.13	4.73	4.44	4.21
50	.100	2.81	2.41	2.20	2.06	1.97	1.90	1.84	1.80
	.050	4.03	3.18	2.79	2.56	2.40	2.29	2.20	2.13
	.025	5.34	3.97	3.39	3.05	2.83	2.67	2.55	2.46
	.010	7.17	5.06	4.20	3.72	3.41	3.19	3.02	2.89
	.001	12.22	7.96	6.34	5.46	4.90	4.51	4.22	4.00

(Continued)

APPENDIX A ● F Distribution (*Continued*)

denominator degrees of freedom	p	9	10	12	15	20	25	30	40	50
		numerator degrees of freedom								
25	.100	1.89	1.87	1.82	1.77	1.72	1.68	1.66	1.63	1.61
	.050	2.28	2.24	2.16	2.09	2.01	1.96	1.92	1.87	1.84
	.025	2.68	2.61	2.51	2.41	2.30	2.23	2.18	2.12	2.08
	.010	3.22	3.13	2.99	2.85	2.70	2.60	2.54	2.45	2.40
	.001	4.71	4.56	4.31	4.06	3.79	3.63	3.52	3.37	3.28
26	.100	1.88	1.86	1.81	1.76	1.71	1.67	1.65	1.61	1.59
	.050	2.27	2.22	2.15	2.07	1.99	1.94	1.90	1.85	1.82
	.025	2.65	2.59	2.49	2.39	2.28	2.21	2.16	2.09	2.05
	.010	3.18	3.09	2.96	2.81	2.66	2.57	2.50	2.42	2.36
	.001	4.64	4.48	4.24	3.99	3.72	3.56	3.44	3.30	3.21
27	.100	1.87	1.85	1.80	1.75	1.70	1.66	1.64	1.60	1.58
	.050	2.25	2.20	2.13	2.06	1.97	1.92	1.88	1.84	1.81
	.025	2.63	2.57	2.47	2.36	2.25	2.18	2.13	2.07	2.03
	.010	3.15	3.06	2.93	2.78	2.63	2.54	2.47	2.38	2.33
	.001	4.57	4.41	4.17	3.92	3.66	3.49	3.38	3.23	3.14
28	.100	1.87	1.84	1.79	1.74	1.69	1.65	1.63	1.59	1.57
	.050	2.24	2.19	2.12	2.04	1.96	1.91	1.87	1.82	1.79
	.025	2.61	2.55	2.45	2.34	2.23	2.16	2.11	2.05	2.01
	.010	3.12	3.03	2.90	2.75	2.60	2.51	2.44	2.35	2.30
	.001	4.50	4.35	4.11	3.86	3.60	3.43	3.32	3.18	3.09
29	.100	1.86	1.83	1.78	1.73	1.68	1.64	1.62	1.58	1.56
	.050	2.22	2.18	2.10	2.03	1.94	1.89	1.85	1.81	1.77
	.025	2.59	2.53	2.43	2.32	2.21	2.14	2.09	2.03	1.99
	.010	3.09	3.00	2.87	2.73	2.57	2.48	2.41	2.33	2.27
	.001	4.45	4.29	4.05	3.80	3.54	3.38	3.27	3.12	3.03
30	.100	1.85	1.82	1.77	1.72	1.67	1.63	1.61	1.57	1.55
	.050	2.21	2.16	2.09	2.01	1.93	1.88	1.84	1.79	1.76
	.025	2.57	2.51	2.41	2.31	2.20	2.12	2.07	2.01	1.97
	.010	3.07	2.98	2.84	2.70	2.55	2.45	2.39	2.30	2.25
	.001	4.39	4.24	4.00	3.75	3.49	3.33	3.22	3.07	2.98
40	.100	1.79	1.76	1.71	1.66	1.61	1.57	1.54	1.51	1.48
	.050	2.12	2.08	2.00	1.92	1.84	1.78	1.74	1.69	1.66
	.025	2.45	2.39	2.29	2.18	2.07	1.99	1.94	1.88	1.83
	.010	2.89	2.80	2.66	2.52	2.37	2.27	2.20	2.11	2.06
	.001	4.02	3.87	3.64	3.40	3.14	2.98	2.87	2.73	2.64
50	.100	1.76	1.73	1.68	1.63	1.57	1.53	1.50	1.46	1.44
	.050	2.07	2.03	1.95	1.87	1.78	1.73	1.69	1.63	1.60
	.025	2.38	2.32	2.22	2.11	1.99	1.92	1.87	1.80	1.75
	.010	2.78	2.70	2.56	2.42	2.27	2.17	2.10	2.01	1.95
	.001	3.82	3.67	3.44	3.20	2.95	2.79	2.68	2.53	2.44

denominator degrees of freedom		p	1	2	3	4	5	6	7	8
			numerator degrees of freedom							
	60	.100	2.79	2.39	2.18	2.04	1.95	1.87	1.82	1.77
		.050	4.00	3.15	2.76	2.53	2.37	2.25	2.17	2.10
		.025	5.29	3.93	3.34	3.01	2.79	2.63	2.51	2.41
		.010	7.08	4.98	4.13	3.65	3.34	3.12	2.95	2.82
		.001	11.97	7.77	6.17	5.31	4.76	4.37	4.09	3.86
	100	.100	2.76	2.36	2.14	2.00	1.91	1.83	1.78	1.73
		.050	3.94	3.09	2.70	2.46	2.31	2.19	2.10	2.03
		.025	5.18	3.83	3.25	2.92	2.70	2.54	2.42	2.32
		.010	6.90	4.82	3.98	3.51	3.21	2.99	2.82	2.69
		.001	11.50	7.41	5.86	5.02	4.48	4.11	3.83	3.61
	200	.100	2.73	2.33	2.11	1.97	1.88	1.80	1.75	1.70
		.050	3.89	3.04	2.65	2.42	2.26	2.14	2.06	1.98
		.025	5.10	3.76	3.18	2.85	2.63	2.47	2.35	2.26
		.010	6.76	4.71	3.88	3.41	3.11	2.89	2.73	2.60
		.001	11.15	7.15	5.63	4.81	4.29	3.92	3.65	3.43
	1000	.100	2.71	2.31	2.09	1.95	1.85	1.78	1.72	1.68
		.050	3.85	3.00	2.61	2.38	2.22	2.11	2.02	1.95
		.025	5.04	3.70	3.13	2.80	2.58	2.42	2.30	2.20
		.010	6.66	4.63	3.80	3.34	3.04	2.82	2.66	2.53
		.001	10.89	6.96	5.46	4.65	4.14	3.78	3.51	3.30

(Continued)

APPENDIX A ● F Distribution (*Continued*)

denominator degrees of freedom	p	9	10	12	15	20	25	30	40	50
		numerator degrees of freedom								
60	.100	1.74	1.71	1.66	1.60	1.54	1.50	1.48	1.44	1.41
	.050	2.04	1.99	1.92	1.84	1.75	1.69	1.65	1.59	1.56
	.025	2.33	2.27	2.17	2.06	1.94	1.87	1.82	1.74	1.70
	.010	2.72	2.63	2.50	2.35	2.20	2.10	2.03	1.94	1.88
	.001	3.69	3.54	3.32	3.08	2.83	2.67	2.55	2.41	2.32
100	.100	1.69	1.66	1.61	1.56	1.49	1.45	1.42	1.38	1.35
	.050	1.97	1.93	1.85	1.77	1.68	1.62	1.57	1.52	1.48
	.025	2.24	2.18	2.08	1.97	1.85	1.77	1.71	1.64	1.59
	.010	2.59	2.50	2.37	2.22	2.07	1.97	1.89	1.80	1.74
	.001	3.44	3.30	3.07	2.84	2.59	2.43	2.32	2.17	2.08
200	.100	1.66	1.63	1.58	1.52	1.46	1.41	1.38	1.34	1.31
	.050	1.93	1.88	1.80	1.72	1.62	1.56	1.52	1.46	1.41
	.025	2.18	2.11	2.01	1.90	1.78	1.70	1.64	1.56	1.51
	.010	2.50	2.41	2.27	2.13	1.97	1.87	1.79	1.69	1.63
	.001	3.26	3.12	2.90	2.67	2.42	2.26	2.15	2.00	1.90
1000	.100	1.64	1.61	1.55	1.49	1.43	1.38	1.35	1.30	1.27
	.050	1.89	1.84	1.76	1.68	1.58	1.52	1.47	1.41	1.36
	.025	2.13	2.06	1.96	1.85	1.72	1.64	1.58	1.50	1.45
	.010	2.43	2.34	2.20	2.06	1.90	1.79	1.72	1.61	1.54
	.001	3.13	2.99	2.77	2.54	2.30	2.14	2.02	1.87	1.77

APPENDIX B ● Chi-Square Distribution

	p				
df	.25	.10	.05	.025	.01
1	1.323	2.706	3.841	5.024	6.635
2	2.773	4.605	5.991	7.378	9.210
3	4.108	6.251	7.815	9.348	11.345
4	5.385	7.779	9.488	11.143	13.277
5	6.626	9.236	11.071	12.833	15.086
6	7.841	10.645	12.592	14.449	16.812
7	9.037	12.017	14.067	16.013	18.475
8	10.219	13.362	15.507	17.535	20.090
9	11.389	14.684	16.919	19.023	21.666
10	12.549	15.987	18.307	20.483	23.209
11	13.701	17.275	19.675	21.920	24.725
12	14.845	18.549	21.026	23.337	26.217
13	15.984	19.812	22.362	24.736	27.688
14	17.117	21.064	23.685	26.119	29.141
15	18.245	22.307	24.996	27.488	30.578
16	19.369	23.542	26.296	28.845	32.000
17	20.489	24.769	27.587	30.191	33.409
18	21.605	25.989	28.869	31.526	34.805
19	22.718	27.204	30.144	32.852	36.191
20	23.828	28.412	31.410	34.170	37.566
21	24.935	29.615	32.671	35.479	38.932
22	26.039	30.813	33.924	36.781	40.289
23	27.141	32.007	35.172	38.076	41.638
24	28.241	33.196	36.415	39.364	42.980
25	29.339	34.382	37.652	40.646	44.314
26	30.435	35.563	38.885	41.923	45.642
27	31.528	36.741	40.113	43.194	46.963
28	32.620	37.916	41.337	44.461	48.278
29	33.711	39.087	42.557	45.722	49.588
30	34.800	40.256	43.773	46.979	50.892
31	35.887	41.422	44.985	48.232	52.191
32	36.973	42.585	46.194	49.480	53.486
33	38.058	43.745	47.400	50.725	54.776
34	39.141	44.903	48.602	51.966	56.061
35	40.223	46.059	49.802	53.203	57.342
36	41.304	47.212	50.998	54.437	58.619
37	42.383	48.363	52.192	55.668	59.892
38	43.462	49.513	53.384	56.896	61.162
39	44.539	50.660	54.572	58.120	62.428
40	45.616	51.805	55.758	59.342	63.691
60	66.981	74.397	79.082	83.298	88.379
100	109.141	118.498	124.342	129.561	135.807

• Glossary •

Additive Difference A difference or change expressed in terms of simple addition or subtraction, for instance a 10-unit increase, meaning adding 10 units, or a 12-unit decrease, meaning subtracting 12 units.

Alternative Hypothesis (H_1) In hypothesis testing, an expression of what some population characteristic may instead be if the current belief or knowledge about it (expressed in the null hypothesis) is incorrect.

Analysis of Variance (ANOVA) Statistical analysis to test the null hypothesis that the mean of some variable Y is the same in each of several group populations.

Antilog A function that undoes a log transformation by raising the base to a particular power; for a base c and a number d, $c^{\log d} = d$, so that raising c to the power log d returns the original number d.

Autocorrelation A violation of one of multiple regression's underlying technical assumptions, with the result that residuals are in some way correlated among observations. Most common for time-series data (temporal autocorrelation), but also could happen when cross-sectional units are nearby, as in neighborhood-level data (spatial autocorrelation).

Base (for Logarithm) The number being raised to some power in the log operation; usually chosen to be 10 (for the standard log) or e (for the natural log).

Base 10 (for Logarithm) Base for the standard log (not natural log); in that case the antilog operation is $10^{\log d} = d$.

Base e (for Logarithm) See *Natural Log.*

Bivariate Regression See *Simple (or Bivariate) Linear Regression.*

Categorical Variable A variable for which values indicate each case's type or category, rather than reporting a numerical measurement.

Causal Inference Methods that attempt to draw conclusions about causal relationships between variables, as in a true experimental design, but from observational (nonexperimental) data; usual regression approaches do not typically allow this sort of causal interpretation.

Centering A transformation of a variable in which the mean is subtracted from each case's value, so that the variable is expressed as deviations from the mean, rather than its raw values.

Chi-square Statistic In general, a test statistic used in a wide variety of hypothesis testing situations; as applied in this book, a test statistic used to compare nested models estimated by maximum likelihood (logit, Poisson, etc.), in order to assess whether the more complicated model (with more X's) sufficiently improves on the less complicated model (with fewer X's) to justify using those additional independent variables; calculated as $(-2 \times \text{log-likelihood}_0) - (-2 \times \text{log-likelihood}_1)$; see also *Chi-square Test.*

Chi-square Test A statistical test involving the calculation of a chi-square statistic and a p-value obtained from a chi-square table; in this book, the chi-square test is mostly used in the context of comparison of nested models that have been estimated by maximum likelihood (logistic regression, Poisson regression, etc.); see also *Chi-square Statistic.*

Classification Table For a logit model, a 2×2 table classifying the sample cases according to whether (a) a case's estimated $P\{Y = 1\}$ from the model was above or below 0.50 and (b) the actual Y for that case was 0 or 1; the classification table indicates the proportion of the sample cases in which the model seemed to "get it right" (by estimating $P\{Y = 1\} > 0.50$ for a case that actually had $Y = 1$, or estimating $P\{Y = 1\} < 0.50$ for a case that actually had $Y = 0$).

Coefficient An independent variable's b_i in a regression equation, multiplied by the value of the independent variable when predicting the dependent variable; see also *Slope (b).*

Collinearity A situation in which strong relationships among the independent variables inhibit the ability of our regression analysis to distinguish effects on the dependent variable, increasing uncertainty in the results.

Confidence Interval An interval (range of values) for an unknown value in the population that takes the general form of a sample estimate plus or minus a "margin of error" for that estimate; the associated confidence level indicates how confident we are that the interval contains the actual population value.

Control Variable An independent variable in the regression model that does not have a central role in the theory being examined but still may be important as a predictor of Y and due to its relationship with independent variables that are central to the theory.

Correlation (r_{xy}) A measure that can range from –1 to 1 and indicates the direction (positive or negative) and strength (where stronger means closer to a straight line) of a linear relationship between two variables; in this book, Pearson's correlation.

Count Variable A variable that can take only integer (whole number) values and cannot be negative.

Cross-Sectional Analysis Analysis of data consisting of measurements of variables taken at the same time (or as close to the same time as practical) for all units in the sample; resulting data are a snapshot of the N units in the sample at essentially a single point in time, such as the unemployment rate in the same year for all of the nations in our sample.

Curvilinear Relationship A relationship between X and Y in which the sign of the relationship changes from positive to negative, or vice versa, depending on whether we are focusing on values of X above or below a particular level.

Degrees of Freedom In hypothesis testing, a value needed in order to identify the specific reference distribution (and place in a statistical table) from which to obtain the p-value; usually takes the generic form of [(number of observations) – (number of parameters estimated in the model)] and is often involved in variance calculation (such as the variance of error terms in least squares regression).

Dependent Variable (Y) The outcome that we are attempting to explain or predict via the independent variables (X) in our regression analysis.

Descriptive Statistics Statistics that summarize fundamental information on the variables measured for cases (units of analysis) in a sample, such as means, standard deviations, ranges, and, for categorical variables, proportions.

Deviance A statistic, used here for count data, which reflects a –2 × log-likelihood comparison between the model being analyzed and the "saturated" model in which every case's count is fit exactly (that is, Y is equal to $\mu^{(Y)}$, its expected value under the model); used to assess model fit.

Dummy Variable A special type of numerical variable, in which information on category membership from a categorical variable has been transformed into a numerical form; takes on values 0 and 1.

Ecological Fallacy An inappropriate attempt to address individual-level research questions or draw individual-level conclusions via aggregate-level data; similar concerns apply to the reverse situation.

Effect An expression of the relationship between an independent variable and the dependent variable in a regression analysis; see also *Slope*; used as short-hand language that does not necessarily imply a causal relationship; see also *Causal Inference*.

Errors See *Residual (Error)*.

Event History Analysis See *Survival Analysis*.

Exhaustive A set of categories in which every case belongs to at least one category.

Exposure For log-rate models, the exposure is typically the size of the population at risk for the event being counted in the dependent variable; in the equation for a log-rate model, the exposure is logged, and referred to as the offset.

Frequency Distribution A table that typically reports the number and percentage of sample cases taking each observed value, or falling into each category, of a variable.

F-Statistic A test statistic used here for comparison of nested regression models to assess whether the improvement in R^2 or sum of squared errors (SSE) in the more complicated model (with more X's) over the less complicated model (with fewer X's) is sufficient to justify the additional X's; calculated as $F = \dfrac{(SSE_0 - SSE_1)/s}{SSE_1/df_1}$ or $F = \dfrac{(R_1^2 - R_0^2)/s}{(1 - R_1^2)/df_1}$.

F-Test A statistical test involving calculation of an F-statistic and a p-value obtained from an F-table; here used in the context of nested model comparison, often for assessing statistical significance of a set of dummy variables.

Hypothesis Test Test that attempts to decide which of two distinct propositions, the *null hypothesis* (denoted H_0) and the *alternative hypothesis* (H_1), about some aspect of the population seems more plausible in light of the sample data.

Independent Variable (X) A factor that we believe will help predict the value of the dependent variable (Y).

Interaction A situation involving a pair (or more) of independent variables in which the effect of one independent variable on the dependent variable changes depending on the value (or category) of another independent variable.

Interaction Terms/Variables Variables that express interaction between independent variables in the regression; constructed by multiplying the independent variables involved in the interaction.

Intercept (a) A part of the regression equation that is used to calculate the predicted dependent variable Y but is not a coefficient on an independent variable X; also the predicted value of Y when all independent variables equal zero.

Least Squares Principle The idea that the numerical values of a and the b's in the regression equation should be chosen so as to minimize ("least") the sum of squared residuals ("squares").

Level of Analysis A characterization of the nature of the units in the sample used for a social research project; units are typically considered either individual-level (such as individual people) or aggregate-level (collections of individuals, such as cities, states, or nations).

Linear Probability Model A linear regression model for a dichotomous (two-category) dependent variable in which effects of independent variables are interpreted as effects on $P\{Y = 1\}$, rather than as effects on Y itself.

Linear Relationship A relationship between X and Y that can be represented (or approximated) by a straight line predicting Y from X.

Log-Likelihood Likelihood is essentially the probability of obtaining the observed sample data on Y if we assume that a particular set of b's gives the correct values in the model equation; log-likelihood is the natural log of the likelihood.

Log Odds For a logit model, the log of odds that Y = 1; that is, $\log \frac{P\{Y = 1\}}{1 - P\{Y = 1\}}$.

Log-Rate Model A version of regression for a count dependent variable that examines rates derived from counts, such as a crime rate based on the number of crimes divided by the population; this model recognizes that Y is a count while also recognizing that, to be meaningful, the analysis must focus on the rate rather than on the raw count.

Logarithmic (log) Transformation Transformation of a variable via the logarithm function so that additive changes in the transformed variable correspond to multiplicative changes in the original variable.

Logging See *Logarithmic (log) Transformation.*

Logistic (Logit) Regression Regression model involving a dichotomous (two-category) dependent variable, constructed as a linear model for the log odds (logit) that Y equals one; the logit model implies that $P\{Y = 1\} = \frac{e^L}{1 + e^L}$, where L represents the logit indicated by the regression model.

Logit See *Log Odds.*

Maximum Likelihood For the models discussed here, an approach used to estimate the intercept and coefficients, in which these values are chosen so as to maximize (make as large as possible) the probability with which the observed sample data would occur; see also *Log-likelihood.*

Mean ($\overline{X}$) A variable's average in a sample, calculated by adding up the variable's values across all of the cases, and dividing by N, the number of cases; $\overline{X} = \frac{\Sigma X}{N}$.

Median A variable's middle value in a sample; after ordering the sample values of a variable from lowest to highest, the middle value (or, for an even number of cases, a value between the two middle values, usually their average).

Moderation Another term for interaction; see also *Interaction.*

Moderator Variable See *Interaction Terms/Variables.*

Multilevel Models Regression models in which data come from more than one level of analysis, such as an analysis of a person-level outcome for which city (aggregate-level) characteristics are among the independent variables.

Multinomial Logistic (logit) Regression Regression model for a categorical dependent variable that has three or more nonordered or nominal categories, such as "Democrat," "Republican," and "Independent."

Multiple Linear Regression An analytic approach that assumes linear (straight-line) relationships between multiple X's and Y; represented by the equation $\hat{Y} = a + b_1 X_1 + b_2 X_2 + b_3 X_3$, where $\hat{Y}$ represents the predicted value of Y, a represents the Y-intercept, b_1 represents the effect (slope) of X_1 on Y, holding the other X's (X_2 and X_3) constant, b_2 represents the effect (slope) of X_2 on Y, holding the other X's (X_1 and X_3) constant, and b_3 represents the effect (slope) of X_3 on Y, holding the other X's (X_1 and X_2) constant; values of a and the b's are obtained by least squares. See also *Least Squares Principle.*

Multiplicative Difference A difference or change in a variable's value that is expressed as a percentage or multiplier; for instance, a 20% increase or, equivalently, multiplication by 1.20.

Multivariate Linear Regression See *Multiple Linear Regression.*

Mutually Exclusive A set of categories in which no case can fall into more than one category.

Naïve Model A regression model that includes no independent variables and predicts the same value of the dependent variable for all cases; in linear regression, the naïve model is $\hat{Y} = a$ and predicts the mean ($\bar{Y}$) for all cases.

Natural Log Logarithm using base e, where e is defined for its first 12 decimal places as e = 2.718281828459. For base e, the antilog operation is $e^{\log d} = d$.

Negative Binomial Regression A regression model for a count dependent variable that extends Poisson regression to account for overdispersion; see also *Poisson Regression* and *Overdispersion.*

Nested Models A situation involving two models for the same dependent variable, in which one model (the larger model) includes all the variables in the other (the smaller model), plus some more.

Nonlinear Relationship A relationship between X and Y that is represented by some curve rather than a straight line.

Null Hypothesis (H_0) In hypothesis testing, a statement expressing current belief or knowledge about some population characteristic.

Odds The odds of {Y = 1} compare P{Y = 1} to 1 − P{Y = 1}, defined as $\frac{P\{Y = 1\}}{1 - P\{Y = 1\}}$; used to construct logit models for categorical dependent variables.

Odds Ratio For an independent variable in a logit model, the estimated multiplicative change in the odds of {Y = 1} when the independent variable increases by one unit, holding all other independent variables constant.

Offset See *Exposure.*

One-sided Hypothesis Test A hypothesis test in which the alternative hypothesis (H_1) states that conceivable values of population lie only above, or only below, the value indicated by the null hypothesis.

Ordinal (ordered) Logistic Regression Regression model involving a categorical dependent variable that has three or more ordered categories such as "strongly disagree," "disagree," "neutral," "agree," and "strongly agree."

Overdispersion For a count dependent variable in a Poisson regression model, a situation in which there is more variability in the observed counts than would be implied by the properties of the Poisson distribution; often addressed by use of negative binomial regression.

Parabola A curve obtained from a mathematical function of the form $Y = d + eX + fX^2$; the parabola's curve is roughly U- or upside-down U-shaped, with interest often focusing on the X value corresponding to the bottom of the U or the top of the upside-down U.

Poisson Regression Regression analysis for a count dependent variable that can only take integer (whole number) values and cannot be negative; the Poisson model is for the expected (mean) value of Y for a case and assumes that the case's observed value of Y follows a Poisson distribution, in which the mean equals the variance.

Population A group that includes all the objects or subjects that a researcher is interested in studying, such as the set of all members of an organization or the set of all American cities with 100,000 or more residents.

Principal Components An approach in which the information in a set of variables, possibly highly correlated, is re-expressed in a set of completely uncorrelated new variables (components); in the ideal case, a relatively small number of components can express almost all of the information in the original set of variables.

Probability Chance of an event occurring; a probability cannot be negative and cannot exceed 1.

Probit Regression Regression model involving a dichotomous (two-category) dependent variable; differs from logit in the way that it models the probability that Y = 1 through the standard normal distribution but typically results in similar conclusions about the research questions.

Pseudo-R^2 A measure of a model's fit to the sample data for models such as logit whose estimation is not based on the least squares principle; attempts to mimic the interpretation of R^2 in least squares regression.

p-value In hypothesis testing, the probability, assuming that the null hypothesis is correct, of obtaining a sample value that is as discrepant (or more) from the hypothesized value as was the statistic that we calculated in our sample; a smaller p-value pushes us more toward rejecting H_0.

R-squared (R^2) The main indicator of the fit (how close the predicted and actual values of the dependent variable Y are for the cases in the sample) of a multiple regression model; R^2 ranges between 0 and 1, with values close to 1 indicating that the multiple regression is predicting Y well in the sample (good fit) and values close to 0 indicating that the multiple regression is not predicting Y well in the sample (poor fit); also described as the proportion of variance (or variation) in the dependent variable that is explained, or accounted for, by the set of independent variables (X's) in the multiple regression.

Range In descriptive statistics, the difference between the minimum and maximum values of a variable across all cases in the sample.

Real-Life Significance Evaluation of whether the size of a statistically significant effect of an independent variable is large enough to be meaningful in terms of the actual social setting for the research; this evaluation is made considering both b and the sample descriptive statistics for the independent and dependent variable; also called Real-World or *Substantive Significance.*

Reference Category When constructing dummy variables for a categorical independent variable, the category that is left out from the set of dummy variables used in the regression; forms the basis for interpreting a dummy variable's b as a difference between categories in the predicted Y.

Regression Coefficient An alternative term for b in the regression equation; in calculating a predicted value of Y, the coefficients multiply the values of the independent variables (the X's).

Regression Diagnostics Supplementary analyses that investigate the context in which the main regression analysis takes place, such as identification of outliers or influential cases and their impact on the main results of the regression.

Regression Line In a scatterplot of X and Y values for cases in a sample, the best straight-line summary of the relationship between X and Y; the regression equation relating the predicted Y to X; see *Simple (or Bivariate) Linear Regression.*

Regression Model A term for the prediction of the dependent variable (Y) via the independent variables (X's), as expressed in the regression equation; see also *Multiple Linear Regression.*

Residual (Error) In simple or multiple regression, the difference between the actual and predicted values of Y for a case in the sample.

Sample A subset of objects or subjects drawn from a larger population, forming the set of cases used for an analysis.

Simple (or Bivariate) Linear Regression Finds the straight-line that best summarizes the relationship between X and Y in a scatterplot from a sample; the equation for simple linear regression is $\hat{Y} = a + bX$, where $\hat{Y}$ represents the predicted value of Y, a represents the Y-intercept, and b represents the slope, effect, or coefficient (the change in $\hat{Y}$ per unit change in X).

Slope (b) In a regression equation, b indicates the change in $\hat{Y}$ per unit change in that X (controlling for other X's in the model in multiple regression); in simple regression, this is the slope of the regression line; *effect* and *coefficient* also refer to the same quantity.

Standard Deviation (s) In descriptive statistics, square root of a variable's variance; represents a typical deviation from the mean for that variable.

Standard Error of b An estimate of sampling variability in b, giving a sense of how much b might vary from one sample to another that we could have drawn from the population; same idea as standard error of the mean but for variability in a regression coefficient rather than in a sample mean.

Standard Error of the Mean A numerical measure that represents estimated sampling variability or variation across possible samples from the population in a variable's mean; typically estimated by $\frac{s}{\sqrt{N}}$.

Standardized Coefficient A regression coefficient expressed in terms of the sample standard deviations, rather than the original units, of the variables; indicates how many standard deviations the predicted Y increases or decreases as X increases by one standard deviation.

Statistical Significance In regression, a term for the situation in which a variable's true effect on Y is not zero, due to our rejecting the null hypothesis (H_0) of $\beta = 0$ in favor of the alternative hypothesis (H_1) of $\beta \neq 0$; in such an instance, the sample leads us to believe that there really is some nonzero effect of X on Y in the entire population.

Structural Equation Models Complex models that can include such features as observed indicators of latent (unobserved) variables, regression-like relationships among latent variables, and disturbance terms associated with different variables; these models provide a modern approach to traditional path analysis.

Substantive Significance See *Real-Life Significance.*

Sum of Squared Errors (SSE) For a regression analysis, summation of the squared errors or residuals (differences between the actual and predicted Y) over all cases in the sample; calculated as $\Sigma(Y - \hat{Y})^2$.

Sum of Squared Residuals See *Sum of Squared Errors (SSE).*

Survival Analysis Regression analysis for dependent variables that indicate durations, such as length of time until an event takes place; designed to account for censoring, in which the study ends before a case experiences the event of interest.

Test Statistic In hypothesis testing, a statistic calculated from the sample data that assesses how much the observed data departs from what is expected under the null hypothesis; the test statistic is compared to some reference distribution, in a statistical table, in order to obtain the p-value.

Time-Series Analysis The analysis of data consisting of multiple observations on the same case, made at a series of (usually many) time points, such as a nation's annual crime rate; distinguished from cross-sectional data, in which observations are made on many observations at the same time.

True Effect (β) In regression, the effect of an X on Y in the whole population or superpopulation, rather than just in the sample (b).

True Model The regression model for the population with true effects rather than estimates from the sample; the regression equation expressed with β's rather than b's.

t-Statistic A test statistic for a variety of hypothesis tests; in regression, used to test the hypothesis that an independent variable has no effect on Y in the whole population; calculated as $\frac{\text{(sample value} - \text{hypothesized value)}}{\text{standard error}}$.

Two-Sided Hypothesis Test A hypothesis test in which the alternative hypothesis states that conceivable values of the statistic in the population lie both above and below the value hypothesized by the null hypothesis.

Variance (s^2) In descriptive statistics, an average of the squared deviation in the sample of a variable from its mean, calculated as $s^2 = \frac{\Sigma(X - \overline{X})^2}{N}$ or $\frac{\Sigma(X - \overline{X})^2}{N - 1}$.

Zero-Inflated Model In regression for count dependent variables, a model that attempts to account for an excess of zero values for the dependent variable relative to the number of zeros that would be expected under a Poisson or Negative Binomial model.

• References •

Hirschi, Travis. (1969). *Causes of Delinquency.* University of California Press.

Shaw, Clifford R., & McKay, Henry D. (1942). *Juvenile Delinquency and Urban Areas.* University of Chicago Press.

• Index •